ADVANCES IN CARTOGRAPHY

ADVANCES IN CARTOGRAPHY

Edited by
J C MULLER
International Institute for Aerospace Survey and Earth Sciences, Enschede, The Netherlands

Published on behalf of the
INTERNATIONAL CARTOGRAPHIC ASSOCIATION
by
ELSEVIER APPLIED SCIENCE
LONDON and NEW YORK

ELSEVIER SCIENCE PUBLISHERS LTD
Crown House, Linton Road, Barking, Essex IG11 8JU, England

Sole Distributor in the USA and Canada
ELSEVIER SCIENCE PUBLISHING CO., INC
655 Avenue of the Americas, New York, NY 10010, USA

WITH 5 TABLES AND 36 ILLUSTRATIONS

British Library Cataloguing in Publication Data

Advances in cartography.
1. Cartography
I. Müller, J. C. II. International Cartographic
Association
526

ISBN 1-85166-603-6

Library of Congress Cataloging-in-Publication Data

Advances in cartography/edited by J. C. Müller.
p. cm.
Includes bibliographical references and index.
ISBN 1-85166-603-6
1. Cartography. I. Müller, J. C. II. International Cartographic
Association.
GA108.A32 1991
526–dc20

91-11479
CIP

Printed in Great Britain at the University Press, Cambridge.

PRESIDENTIAL FOREWORD

The International Cartographic Association (ICA) was established in 1959 by 13 founding member countries at a meeting held in Bern, Switzerland. Over the years, membership has grown to 62 member countries in 1990. The ICA is the World body for cartography and has as one of its major aims the initiation and coordination for cartographic research through cooperation among cartographers from all member nations. A major mechanism by which this aim is achieved is through the work of the various Commissions and Working Groups of the Association.

This volume on *Advances in Cartography* is an excellent example of international cooperation. The editor Dr Jean-Claude Muller, is a member of the ICA Standing Commission on Advanced Technology, and cartographers from eight ICA member nations have contributed to the volume.

As cartography is poised to enter the 21st century, a volume which looks at the key research issues facing the discipline is most timely. Dr Muller and his colleagues are to be congratulated for this major contribution to the cartographic literature.

D R F Taylor
President
International Cartographic Association

PREFACE

The past decade has seen a tremendous growth in the tools available for cartographers and their application to a diverse range of problems. The rapid development of geographical information systems and their spread from mainframe computers to minicomputers to workstations to personal computers has brought a powerful capability to a much larger community of users. Each new application area poses challenges and problems in implementation and often raises fundamental issues that demand further research. The wide range of research issues addressed in this volume is indicative of the growing importance of cartography in the information age. The efforts of Dr Muller and his colleagues in assessing the current state of research in cartography provide a foundation for continued research and cooperation in the cartographic community.

Dr Muller has performed a great service on behalf of the ICA Standing Commission on Advanced Technology by assembling a distinguished group of scientists to share their knowledge with the community. On behalf of the Commission, a sincere appreciation is due to Dr Müller and his colleagues.

K E Anderson
Chair
Advanced Technology Commission
International Cartographic Association

CONTENTS

LIST OF CONTRIBUTORS

C BOARD
Department of Geography, London School of Economics and Political Science, London, UK

N R CHRISMAN
*Geography DP-*10*, University of Washington, Washington* 98195*, USA*

H CLAUSSEN
*Robert Bosch GmbH, Department MC/ENS*1*, Hildesheim, FRG*

A U FRANK
National Center for Geographic Information and Analysis, University of Maine, Orono, ME 04469*, USA* and *Department for Geoinformation, Technical University of Vienna, Gusshausstrasse* 27-29*, A*-1040 *Vienna, Austria*

M F GOODCHILD
National Center for Geographic Information and Analysis, University of California, Santa Barbara, CA 93106*, USA*

P HAYWOOD
*Ordnance Survey, Romsey Road, Maybush, Southampton SO*9 4*DH, UK*

D M MARK
National Center for Geographic Information and Analysis, Department of Geography, State University of New York at Buffalo, Buffalo, New York 14261*, USA*

D MEDYCKYJ-SCOTT
*Midlands Regional Research Laboratory and Department of Computer Studies, University of Technology, Loughborough, Leicestershire, LE*11 3*TU, UK*

H MOELLERING
Department of Geography 103 *BK, Ohio State University, Columbus, Ohio* 43210*, USA*

J C MULLER
International Institute for Aerospace Survey and Earth Sciences, 350 *Boulevard* 1945, 7500 *AA Enschede, The Netherlands*

W D RASE
Federal Research Institute for Regional Planning, (*BFLR*), *Postfach* 200130, *D*-5300 *Bonn* 2, *FRG*

V B ROBINSON
Institute for Land Information Management, University of Toronto, Mississauga, Ontario, Canada L5L 1*C6*

F SALGÉ
Institut Géographique National, B.P. 68, 94160 *Saint-Mandé, France*

Chapter 1

PROSPECTS FOR AND IMPEDIMENTS AGAINST A NEW CARTOGRAPHY IN THE 1990S

J.C. Muller
ITC, Enschede, The Netherlands

This book is about the cartographic research agenda of the 1990s. On the basis of what has already been accomplished, each author gives his own view, in his own particular area of expertise, of what might still be accomplished or should be done in order to promote further progress in cartography and Geographic Information Systems (GIS). I say cartography and GIS, bearing in mind that research in one area is inseparable from research and development in the other. Maps are essential inputs to and outputs from GISs, and may be looked at as the windows of geographic information. But cartography does not only deal with maps to be seen. A cartographic database, for instance, is another kind of map, a 'virtual' map, which provides a 'materialized view' of the GIS (Frank, Chapter 2).

In order to be able to draw perspectives for future research, one must look at the past. It has become fashionable to talk about a 'new cartography' lately. This was one of the main topics discussed at the last general assembly of the International Cartographic Association (ICA) in Morelia, Mexico, in 1987. New cartography as opposed to old cartography? The question is what is old and what has become new. This antinomic terminology certainly reflects an 'etat d'esprit' upon which I would like to elaborate. Certainly this state of mind becomes obvious if one looks at the titles of books and papers already published. I refer here to the book by Monmonier *Technological Transition in cartography* (1985), the articles published by two former ICA Presidents, 'Revolution in Cartography' by Robinson (1976) and 'The Revolution in Cartography in the 1980s' by Morrison (1989), and the paper of the current president of the German Society for Cartography, 'Do we need a new Cartography?' by Freitag (1987). Although one may hesitate to adopt the word 'revolution' in view of its use and abuse to sensationalize scientific progress, it seems as if we are now living in a permanent state of revolution, both politically and environmentally. Let us look at the historical perspective in order to better appreciate present and prospective changes in cartography.

THE HISTORICAL PERSPECTIVE

What are the engines of progress in cartography? Since ancient times, cartography has been used to fulfil our quest for a better understanding of geographical space and geographical information. Like other documents in textual, numerical or graphical forms, maps have profited from the successive revolutions in hard-copy information technologies, from the invention of rolled-up manuscripts and of paper to the book form, from lithography to xerography and laser printing. There are more paper maps produced today than ever before. But the hard-copy revolutions are only one aspect of the technological changes. The more recent so-called 'soft-copy' revolution, for instance, led the way to the production of ephemeral maps which can be viewed on the screen of a video monitor. Vehicle navigation through visual instruction is a popular application of this technology (Claussen and Mark, Chapter 8). Finally, the digital encoding of spatial information gave us the ability to produce cartographic databases which can be much more diversified in form and structure than a simple digital array resulting from an optically scanned paper map. One would be hard pressed if asked to comment on the forms, dimensions, and relationships of soil entities by looking at a polygon-chain listing or a relational table. And yet those databases are maps in their own right, in the sense that they function operationally as sources of spatial information to be queried and manipulated. But the extension of the notion of cartography beyond the directly visible analogy of space has profound implications. For the first time we are emancipated from the tyranny of the paper map sheet with finite size and depth, although much research still needs to be done towards the efficient management of multilayer, seamless, and scale variable cartographic databases (Chapters 2, 3 and 4). Furthermore, for the first time in the long history of cartography the map is functioning as an abstract algebra paradigm where elements can be added, divided, exponentiated in a GIS tool box for spatial analysis and modelling (Berry 1987). The gentle slope of progress in the first half of this century, characterized by the introduction of stereo-photography, halftone printing, plastic material and scribing (Robinson, 1976) is now replaced by giant steps which probably justify the previous claims for a 'new' cartography.

It is difficult to foresee what the essence of cartography will be ten years from now. What will be the role of maps in the context of hypermedia technology integrating text, image, animation and sound? Will cartographic databases be intelligent enough to provide us with guidance in our spatial queries? One can only speculate about the implication of new technologies such as massive storage, parallel processing, animation software, artificial intelligence programming and synthesized sound production. The purpose of this volume is not to answer those questions, but rather to provide directions as to where we should be going in order to address some of the pressing issues regarding cartography and GIS today. In a backward chaining fashion, I will point out some of the impediments which are believed to obstruct the way towards progress, and let the experts of this volume speak themselves about potential solutions.

IMPEDIMENTS AGAINST A NEW CARTOGRAPHY

There are theoretical impediments, which concern our methodologies in cartographic research, and practical impediments, which are raised in the various fields of application, such as database design and management, standards, production lines and institutions. The distinction between theory and practice is somewhat artificial, but is adopted for convenience of the discussion.

Theoretical Impediments

First, we lack a research agenda with clear priorities. There are too many 'preliminary studies', too many 'initiatives', too many 'prototypes' with not enough validation and developmental work. It is true that scientific research needs redundancy and freedom of pursuit, but the general impression from the recent explosion of conference proceedings is a forest of starting projects with very few trees which stand up and continue to grow. We need more communication, coordination and developmental efforts on the big questions. What are these? Although we probably cannot all agree on a common set of priorities, there is a need for an answer. The scarcity of resources granted by the various R and D agencies would recommend that we spend more time in identifying research priorities.

We need a comprehensive research model that integrates the many technologies involved in the production of geo-information (Figure 1). Such a model could act as a clearing house for research guidance pointing out short and medium term priorities.

Another theoretical impediment is the lack of understanding of the structure of maps, the process that produces them and the lack of a model for map use. We do not know how people perceive space, let alone how they read maps, in spite of some thirty years of research efforts. This raises serious problems for the design of cartographic data structures (Frank, Chapter 2), of spatial query languages (Robinson, Chapter 5), for the communication of data quality (Goodchild, Chapter 6) and the improvement of user interface in a GIS. The recent push towards knowledge based spatial information systems (eg with some front end intelligence to help the user to manipulate information over space with the help of spatial operators) is encouraging; it means new efforts in attempting to formalize the mental and visual cues that we use to perceive and understand spatial relationships. This evolution is analogous to the past evolution from spoken to written word. The latter brought an emphasis on the spelling and grammar, rather than the content of the language. In turn, written words allowed knowledge to be archived which led to greater attention to the definition and use of oral expressions as well as discrepancies between various usages (Parsaye et al. 1989). The danger, of course, is to fall back onto previous, unproductive discussions about cartographic language. Is a cartographic entity a word? Is there such a thing as a cartographic grammar? There have been strong arguments against a pure linguistic approach to map design and map use (Robinson and Petchenik 1976).

Finally, our basic models of cartographic communication will have to be changed, incorporating the social, economic and institutional layers (Chrisman, Chapter 11), as well as the individual knowledge and experience of the map user (Medyckyj-Scott and Board, Chapter 10). Map communication theories and communication models were developed in the 1960s and 1970s in order to better understand the interaction between the map and the map reader. Those attempts led to little practical success, however, due to the variance of the experimental results and/or inability to understand how the brain works. Cognitive cartography research has recently regained the attention of researchers who are involved in the general fields of ergonomics and visualization. It is of paramount importance to understand the way we process information visually and mentally in order to assess the effectiveness of our map products and improve on the techniques of map rendering. A theory of map reading from a cognitive perspective, involving space semantics, perceptual processes, and mental imagery, is urgently needed. Such a theory could contribute to more effective ergonomics in car navigation systems. The building of knowledge-based map design systems could also benefit from this theory, by incorporating rules which are not derived from abstract cartographic principles, but are based on the user's cultural background and needs. A cognitively based cartographic advisor system to help the non-expert to make maps will take some time to develop, which means

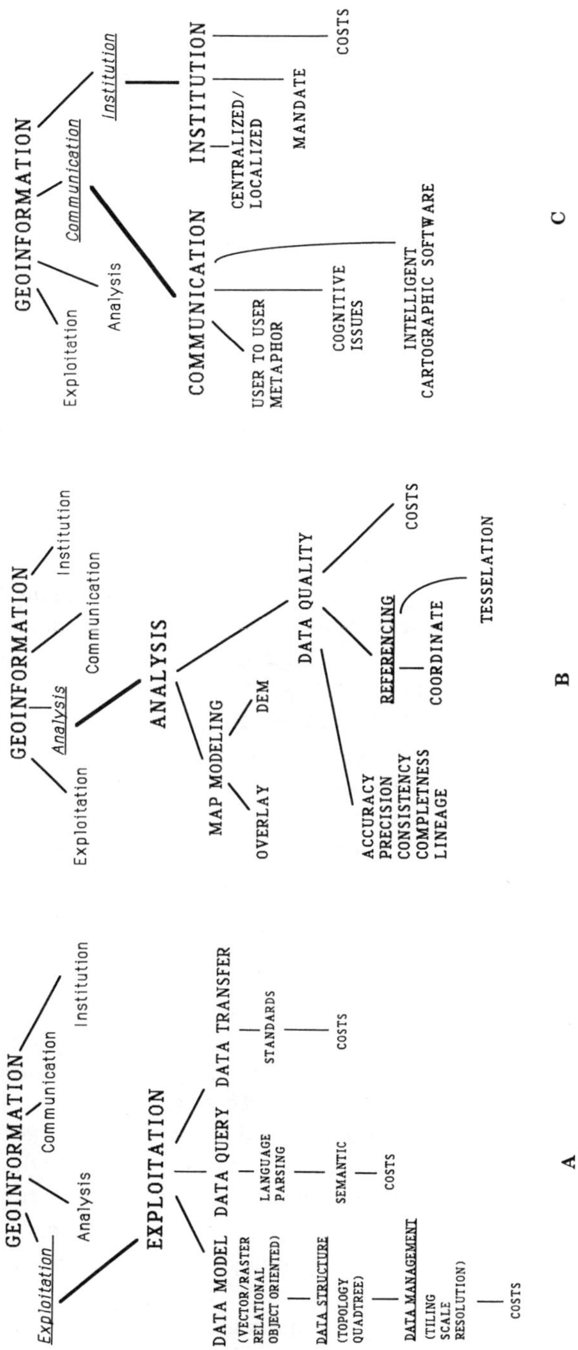
GEOINFORMATION
Exploitation
Analysis
Communication
Institution
EXPLOITATION
DATA MODEL
(VECTOR/RASTER
RELATIONAL
OBJECT ORIENTED)
DATA STRUCTURE
(TOPOLOGY
QUADTREE)
DATA MANAGEMENT
(TILING
SCALE
RESOLUTION)
COSTS
DATA QUERY
LANGUAGE
PARSING
SEMANTIC
COSTS
DATA TRANSFER
STANDARDS
COSTS
A
GEOINFORMATION
Exploitation
Analysis
Communication
Institution
ANALYSIS
MAP MODELING
OVERLAY
DEM
DATA QUALITY
ACCURACY
PRECISION
CONSISTENCY
COMPLETNESS
LINEAGE
REFERENCING
COORDINATE
TESSELATION
COSTS
B
GEOINFORMATION
Exploitation
Analysis
Communication
Institution
COMMUNICATION
USER TO USER
METAPHOR
COGNITIVE
ISSUES
INTELLIGENT
CARTOGRAPHIC SOFTWARE
INSTITUTION
CENTRALIZED/
LOCALIZED
MANDATE
COSTS
C

FIGURE 1

that we can expect GISs and mapping packages to litter our media with more bad cartographic products for a while.

Practical Impediments

For convenience of the discussion, I will distinguish three areas where practical impediments may be identified:

(1) exploitation
(2) control
(3) management

Exploitation

Exploitation must be understood in the wider sense of data access, analysis and interpretation. Hence, the topic includes the design and development of cartographic databases, spatial query languages, and automated map production. Here are the key words encapsulating impediments in those applications.

A major impediment is the slow progress in the organizational aspects of geographical/cartographical databases. Management of data in organization is likely to be the challenge of the future (Haywood, Chapter 4). Models known from a database environment (hierarchical, relational, object oriented) are inefficient at handling geographical information. The conventional approach of separating the geometric and attribute information of an object is not ideal but a collapse of both in one single relational database (location becomes another attribute) is awkward and shows poor performance for data retrieval and manipulation (Salge, Chapter 3). Another impediment is the current tiling of scale-dependent cartographic databases according to the traditional map sheet formats. The lack of spatial continuity may create problems of data integrity at the borders of the tiles and slow down spatial overlay and query operations over several tiles. Furthermore, the map sheet constraint dictates much of our thinking about data management and prevents other products arising beyond the traditional map categories. Since the user of the future will be mainly outside the producer's organization, we need to change our conceptual models in order to respond to the variety of needs coming from the market place. This is particularly relevant in view of the fact that speed and lower cost of digital methods mean that customization and currency of information no longer pose intractable problems. In multimedia hypermaps, for instance, direct imagery can be produced and used to query spatial databases. Furthermore, scale dependency appears to be a major impediment both from a management and scientific point of view. Assume four cartographic individual databases for the corresponding scales 1:20 000, 1:50 000, 1:100 000 and 1:200 000. Each database is partly either a subset or a superset of another database. Hence we have redundancy in storage. This might create problems in consistency since one has to make sure that every update on the 1:20 000 is also reported for the other scales. Since the databases are only partly overlapping, one update on one scale does not automatically affect the other scales. Further, the update of one feature may affect other adjacent features differently at different scales, hence complicated tracing routines may be required which raise the possibilities of inconsistent corrections. Finally, there is the problem that a new scale, say 1:150000, cannot be automatically generated from the existing datasets. This is unfortunate, since physical and human geographic features display characteristics whose scale dependency does not necessarily correspond to the arbitrary discrete scales adopted above.

In the context of digital technology, one would like to go beyond the traditional framework of the paper map series and provide flexibility both through space (by changing the tiling format), and through scale (by changing scale as needed).

Query languages are the key to the access and exploitation of geo-information. How can we allow the user to manipulate data without burdening him with details of data structures and storage? Natural language would be the ideal language for database interaction, reducing to a minimum the inhibition of the GIS user. One of the present impediments is the lack of an effective user interface which incorporates a language for spatial analysis (Robinson, Chapter 5). Spatial Query Language (SpQL) needs more research into spatial semantics (how can we handle the spatial relation 'near' in a meaningful manner?).

Automated map production is probably the area which has shown the most visible achievements in the last twenty years. Not only were we able to automate conventional production lines, but we extended the palette of cartographic products which were otherwise too costly or computationally too intensive to produce. Examples are multivariate maps, cartograms, and three- and four-dimensional representations through animation (Gersmehl 1990). The recent development of Desktop Publishing Systems (DPS) and hypermaps (eg maps linked to textual and numerical databases) promotes automated cartography further. In spite of sophisticated user interface devices, such as pop-up/pop-down menus, windows, icons, pointing devices, acoustic channels, etc, there is little development of the 'back-end' functions of automated cartography, including selection of map types, visual variables, graphic outputs, and support for reproduction and printing (Rase, Chapter 9). An impediment of automated cartography is the lack of 'intelligent' front ends to the cartographic execution programs now available on the market. The absence of a fully or semi-automated generalization facility is another related impediment to automated map production. This is a commodity which will be increasingly in demand. The present strategy where map sheets are digitized and stored at various scales is one practical solution. The notion of a central scale-independent geographic database would be a powerful alternative, but is difficult to justify without a generalization facility for changing resolution and scale.

The next series of impediments discussed belong to control and management.

Control

Data control involves all activities which are related to the monitoring of the quality and integrity of geographic information and the set-up of standards for exchange of cartographic databases.

Quality standards for geographic data are used to cover the following aspects: precision, accuracy, consistency, completeness and lineage. Although accuracy standards are a recognized tradition in the practice of surveying and mapping, those standards are more difficult to formulate or to enforce in the production of geographic information. The impediments here are coming from the lack of research in expressing and handling data error, as well as the multiplicity of agencies and private firms which produce geographic information. GIS technology will likely raise the user's awareness of errors and distortions in spatial data because of the procedures and transformations involved. Operations such as raster/vector conversions and map overlays expose the potential weakness of data much more than did the paper map in the past (Goodchild, Chapter 6). GIS technology affords a degree of precision which is rarely warranted by the accuracy of geographic data (there is no sense in spending money on precision when our methods of sampling and analysis lead to gross inaccuracy). One solution proposed is to adjust precision to accuracy by replacing the conventional coordinate referencing system with some sort of hierarchical tesseral scheme whose level of resolution corresponds to the accuracy of the data (Dutton 1989). This is somewhat analogous to the generalization practice on paper maps, where inaccurate data are portrayed by exaggerated physical marks. Another obstacle to progress in data quality is our poor understanding of the relationship between accuracy and cost. Is it worth the effort to measure and tag the accuracy of a location or an attribute and trace down its effect on complex operations such as successive map overlays? Does this cost less (or more) than the litigation and court expenses which may occur from ignoring the error? There is an increasing pressure on the providers of data to supply information on data quality (Goodchild, Chapter 6). This information is not free, however, and it will be difficult to convince the buyer to pay for it unless there is a better understanding of the cost/benefit relationship.

Transfer of data between agencies, local governments and communities requires consistency in data. Impediments against consistency are the multiplicity of data sources (field survey, existing maps, photography, digital remote sensing, etc) and the multiplicity of agencies working with their own specifications. Often, the specifications are user defined rather than following general standards. Although complete standardization of data accuracy and format is probably unrealistic, proper documentation of the sources, such as level of geodetic control, map scale, resolution of scanner, etc, would help considerably in controlling potential inconsistencies.

A challenge for the surveying and cartographic profession is to provide the guidelines for efficient data exchange and communication. Another challenge, and a more difficult one, is to convince the producer of geo-information to adopt the standards. International transfer standards are developed in various parts of the world. There is a need to evaluate the work which must be carried out for conversion to standards, the feasibility and practicality of interlinking various national databases, as well as the long-term benefits in portability and maintainability of standardized data. A growing impediment against standards is theoretical, however, and refers to the various interpretations or meanings given to spatial data. We need to agree on the semantic in order to provide consistent definitions of data at the source, in the GIS, and in the target products (Moellering, Chapter 7).

Another important area where impediments occur is in management issues of cartographic databases, including the institutions' responsibilities (perceived or real) for producing and managing geo-information.

Management

Abler (1987) identified four kinds of impediment against the adoption of information systems:

(1) technological obstacles
(2) economic constraints
(3) institutional hurdles
(4) cultural barriers

It is common to say that the main impediments against progress in science and development are not technological, but institutional and cultural. That is, we take the technology for granted. There is a complex dialectic relationship between technology and institutions, however, where technology may allow institutions to grow until a point is reached where institutions may be an obstacle to the application of new technologies (Chrisman, Chapter 11). We need to know more about the relationships between technology, institution and management, particularly for the developing countries where technology may be acquired without the scientific and management literacy to make it work. Indeed, the scientific and management literacy required to work with a GIS is as demanding today as was cultural literacy for understanding the Greek philosophers in earlier times.

Another impediment is the confusion regarding mandate and responsibility for data collection and maintenance. Traditionally, national mapping agencies have been responsible for the creation and maintenance of national mapping programmes. Provincial and municipal authorities are now custodians of their own maps as well. Should they be left alone in providing the updating and maintenance of their own data? Do they have the financial means to do so? Is a physically decentralized but conceptually centralized database administration a workable concept (Chrisman, Chapter 11). One must find a formula which ensures a firm control at the central level for data integrity and maintenance with the necessary delegation of power and accountability at the local level as well. Ideally, all levels should cooperate. But one cannot force cooperation; cooperation happens only when there is something to be gained by all parties involved.

As for exploitation and control, management of cartographic data would be made easier if there was a unified theory of spatial data and spatial information usage which would provide some basis for policy and direction. Production managers are reluctant to make organizational changes every time there is a new technological innovation (Haywood, Chapter 4). A recurrent argument in the chapters of this book is the need for good theories to guide the application of rapidly evolving technologies. This argument for good theory is equally applicable for the training of personnel. We know that the hardware has a useful lifespan of about five years; software lasts about ten years; databases remain useful for twenty years or more. What is the lifespan of brainware or orgware? To know this would probably lead to a re-evaluation of our priorities, considering that personnel are the most important of our resources.

CONCLUSIONS

Where will cartography be ten years from now? Will the impediments against a 'new' cartography be removed by the year AD2000? Will there be a post-era to the new cartography? Will new cartographic products — not available today — be sold in the market place? Are the visionary prospects of production systems being installed in high street shops and service stations (Haywood, Chapter 4) merely science fiction? Chrisman (Chapter 11) argues that 'most of the research in scientific cartography has concentrated on re-engineering the procedures, assuming that the products are given'. In other words, the products of today are the expression of past technology, even if we use new tools to produce them. This is certainly true for the electronic bit map, and the metaphor of the evolution of the car, from the horse and carriage, to the horseless carriage and the modern automobile is still valid. We are still partly busy producing horseless carriages, not modern automobiles. It is difficult to define or foresee what kind of products will qualify for being automobiles, however. Does the addition of topology in a spagueti file give a new type of cartographic product? One cannot judge solely on the appearance of a product to make it qualify for the new cartography. The new product will emerge from its ability to afford new kinds of activities which were not performed before. Hence, a topological cartographic file is already a new product, because of the way it can be manipulated to perform Boolean algebra or do network analysis in a GIS. Furthermore, the Digital Elevation Model (DEM) derived from stereo-correlation of SPOT satellite images on a personal computer (Welch 1990) is also a new product, because of its potential modelling capabilities. In contrast, the CRT maps available in Vehicle Navigation Systems or the maps produced with an Apple or DOS Machine remain conventional products, since their main function is one of communication, an old tradition in cartography. On the other hand, conventional map products may lead to a new cartography when operationally integrated with other media, such as text and sound (eg hypermaps which allow querying of text with graphics and vice-versa) or when a new dimension is added (such as time in animated maps for time series analysis and modelling). Therefore, the boundary between 'old' and 'new' cartography is not always clear.

The noticeable phenomenon which has philosophical implications in the trend towards a 'new' cartography is the 'equalizing' effect of technology (Chrisman, Chapter 11). Maps are now produced by the user for the user with the help of commercial micro-based computer packages. The traditional division of labour between map makers and map users is disappearing. This new development results in the production of customized maps which are topically much more specific and flexible than the maps produced by mapping agencies. As a result, the basic models of cartographic communication developed in the 1960s and the 1970s, when the production was monolithic and monopolized by cartographers, are no longer valid. New models will have to incorporate the cartographic naivety of the user, as well as the motivations for making the map in the first place. Individual initiatives coupled with geographical and cartographical illiteracy may result in some kind of chaotic production, which is already apparent on the maps produced by poorly designed cartographic softwares. A helpful hand may be coming from Artificial Intelligence (AI) technology. It must be possible to encapsulate the expert knowledge of a cartographer into a knowledge-based system. The system could be consulted by the user of a cartographic package. The problem here is that academic cartography does not agree with what cartographic expertise is and which cartographic rules must be applied, because there is no universal consensus about what are the ingredients of a good map (a good map may be operationally defined as a map which affords effective usage and communication of spatial information). A bad map is perhaps more easily recognized. This leads me to suggest that we should develop 'negative' Systems which would discourage dangerous practices rather than prescribe mapping choices. This conservative approach will not prevent cartographic mistakes occurring (just as a statistical analysis package does not prevent wrong assumptions being made in running a correlation analysis), but would avoid major blunders (a statistical package does crash when you ask for a regression on qualitative data).

Where will cartography be ten years from now? It is difficult to appreciate the effects of recent technological and theoretical developments on the cartographic discipline. Will cartography remain a discipline of its own? Some doubt 'whether an entire discipline will still exist to make and study [maps] ...; most likely expertise for creating maps will reside within the tool-creating enterprise that is increasingly being referred to as 'spatial data handling' (Petchenik 1988). There is a growing perception that cartography is part of a bigger picture involving other disciplines such as geography and civil engineering and other technologies such as GIS and Remote Sensing. Automated cartography may appear threatening to the professional cartographer, but so did photogrammetry for the land surveyor. The remarkable development in recent years is the increasing awareness of a communality of goals between surveyors, photogrammetrists, cartographers, geographers and other geo information producing disciplines. One would be ill advised to try to protect one's own ground on the basis of tactical and financial concerns, or worse, to lobby the legislatures to set up professional jurisdictions and prevent others from getting involved in what is considered to be one's own territory. Instead, the joining of forces by different disciplines is a requirement for the survival of all, as each discipline is individually incapable of covering the scope of technology and knowledge which is needed for the production of this new commodity called geo information. The contributors of this book coming from different directions is a case in point.

ACKNOWLEDGEMENTS

This research represents part of Research Initiative 3, 'Multiple Representations' and Research Initiative 8, 'Cartographic Expert Systems' of the NCGIA, supported by a grant from the National Science Foundation (SES–88–10917). Support from NSF is gratefully acknowledged.

REFERENCES

Abler R F (1987) The National Science Foundation Center for Geographic Information and Analysis. *International Journal of Geographic Information systems* **1**: 303–26

Berry K J (1987) Computer-assisted map analysis: Potential and Pitfalls. *Photogrammetric, Engineering and Remote Sensing* **53**: 1405–10

Dutton G (1989) Modeling locational uncertainty via hierarchical tesselation. In: Goodchild M F and Gopal S (eds) *Accuracy of Spatial Databases*. New York, Taylor and Francis: 125–40

Freitag U (1987) Do we need a new cartography. *Nachrichten aus dem Karten und Vermessungswesen*, Series II, (46): 51–9

Gersmehl P (1990) Choosing tools: Nine metaphors of four-dimensional cartography. *Cartographic Perspectives*, (5): 3–9

Monmonier M S (1985) *Technological Transition in Cartography*. Madison, University of Wisconsin Press

Morrison J L (1989) The revolution in cartography in the 1980s. In: Rhind D W and Taylor D R F (eds) *Cartography, past, present and future*. Amsterdam, Elsevier ICA Publications

Parsaye K, Chignell M, Khoshafian S, Wong H (1989) *Intelligent Databases*. New York, Wiley and Sons

Petchenik B. B (1988) Afterword. *The American Cartographer* **15**: 321–22

Robinson A H (1976) *Revolution in Cartography. Proceedings ACSM*: 76–142

Robinson A H Petchenik B B (1976) *The Nature of Maps: Essays Towards Understanding Maps and Mapping*. London, University of Chicago Press

Welch R (1990) Three dimensional terrain modeling for GIS application.*GIS World*, October issue: 26–30

Chapter 2

DESIGN OF CARTOGRAPHIC DATABASES

Andrew U. Frank
National Center for Geographic Information and Analysis, University of Maine, Orono, ME 04469, USA *and* Department for Geoinformation, Technical University of Vienna, Gusshausstrasse 27-29, A-1040 Vienna, Austria

INTRODUCTION

Computers have been found to be very useful tools for helping with the production of maps. The contents of a map sheet can be stored in computer memory and changes easily introduced. An automatic plotter can then be used to redraw the map. Today, many organizations that have to maintain collections of maps, from the traditional mapping agencies to public utilities, use automated mapping procedures in one form or another. The organization of storage of cartographic data in the most effective manner becomes then a topic that needs attention (Buchmann et al. 1990), and criteria for optimization are necessary.

Cartographic data are voluminous and therefore storage should be compact. Compact storage not only saves money as less storage capacity needs to be bought but it also reduces processing time. This argument is less important today, as the reduction in cost for storage devices, primarily for magnetic disk devices, has been dramatic in recent years. In 1990, hard disk devices cost around US$15 per megabyte[1] and prices are falling further. Disk capacity is around 1 Gigabyte. The new optical disc technologies offer very inexpensive readers — prices are less than US$1000 and should fall further in the next few years. The medium is inexpensive, the production of a 600 Megabyte disk (similar in size to the Compact Discs used for music) is in the US$10 range and a number of distributors offer cartographic data in this format. It is apparent that storage space and cost of the necessary devices is not a dominant concern any more and we need only worry about orders of magnitudes. It has been repeatedly observed that the cost of storage is dropping faster than most agencies can collect the data; ie the cost of buying all the storage devices necessary to store today's data is no more than the cost of the devices necessary a few years ago when data volumes where much less, and it seems likely that costs will be the same for storing the vastly increased amounts of data in the future. Storage cost per organization seems constant and independent of the increase in data volume due to cost reduction by advances in technology.

The storage size is also relevant for processing, as many operations require, in first approximation, time proportional to the amount of data being processed (Samet 1989*a*: 37). Fortunately, advances in processing speed have been very fast in the last few years, currently doubling in a single year, making this consideration less important. An exception is the delay in accessing data stored on disk, which has remained relatively constant during the past decade and will remain slow. Access to data on a disk takes about 20 to 50 milliseconds, that is about 10^6 times slower than accessing data stored in the main memory.

In conclusion, technical reasons do not limit the size of databases anymore. Data compression — transparent to the user — remains an interesting research topic, but saving storage space should not be a primary driving force in the design of a cartographic database.

On the other hand, we observe that collecting and maintaining the data is a major cost. Data collection is the primary investment and thus it is imperative that the data can be used effectively and for a long time. Our attention must therefore concentrate on methods of simplifying and expanding the use of the data, and technology must be used to facilitate this goal. Experience over the last decade shows that technological advances are very fast. Hardware progress especially is generally faster than expected. Software advances are much slower, and to make matters worse, most people overestimate what can be done and how soon it will be ready for use — the so-called 'software crisis' (Traub 1989). Least progress is made in understanding the organizational framework in which a cartographic database is constructed and used. It has been shown that introducing new technology in an existing organization is difficult and slow. The technical solution should be adapted to the organizational requirements; we should not try to force the organization into a technical solution that does not fit. Management of data in organizations is likely to be the challenge of the future (Egenhofer and Frank 1990, Frank 1980, Haywood, Chapter 4).

In consequence, this chapter will concentrate on the logical issues that are likely to affect cartographic databases and how they are used within organizations for a number of years to come. It seems important to avoid spending time and effort on issues that will be resolved automatically by more powerful hardware, and to concentrate on the topics that will persist and need solutions based on a cartographer's understanding of the goals, uses, and limitations of cartography.

We first attempt to clarify the notion of a cartographic database and compare it with the related term 'geographic information system'. We then discuss different concepts for modelling a cartographic database, specifically understanding a map as an image, as the result of drawing instructions as a structured set of symbols or as a representation of map features. In the fourth section, we consider the stages of the mapping process. In the following sections, we investigate the data structure and database requirements for linking a GIS and a cartographic database and then discuss the simpler case of constructing multi-scale cartographic databases that can deal with map series. We conclude with two sets of recommendations, one for practitioners and one with open research questions.

WHAT IS A CARTOGRAPHIC DATABASE?

In the last few years a number of cartographic institutions have built databases or otherwise organized the collection of cartographic data they use to produce maps. Such collections are probably properly called cartographic databases. It must then be asked how they differ from geographic information systems, a term currently in wide used (Frank 1980). In this section, we try to give reasonable definitions for these terms. This is difficult in a field that is very young, progressing rapidly, and that seems to follow Humpty Dumpty's rule 'When I use a word, it means just what I choose it to mean — neither more nor less' (Carroll 1960: 269), often without regard for etymology, taxonomic clarity or the established use of terminology in related fields.

Databases and Database Management Systems

In very general terms, a database is a collection of data together with the methods to manage these data. Just a collection of data without the organization and methods for access, etc, is not justly called a database. A database consists of the software needed for storing and retrieving the data as well as the data which have been stored (Figure 1) (Zehnder 1981, Codd 1982, Date 1986).

The programming of the database software is necessarily very complex in order to respond to all the demands of storing and retrieving data in a safe and secure manner. The computer industry has produced Database Management Systems (DBMS) which are general purpose and fulfil generic database requirements. They are adapted by the database manager to a specific use, ie to manage a specific data collection.

A database management system should provide the following functions (Frank 1988):

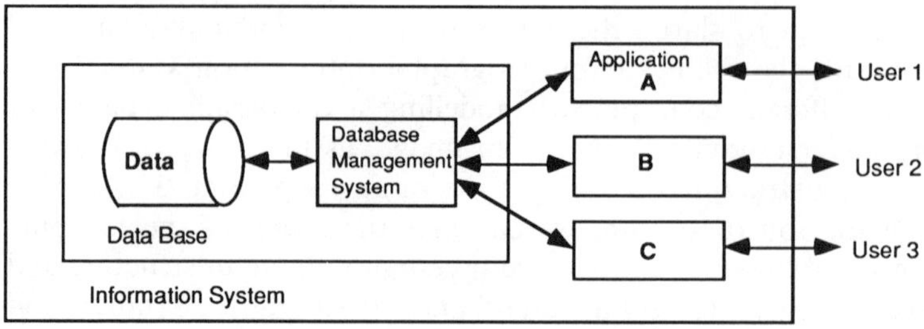

FIGURE 1 Database concept (Source: Frank 1988)

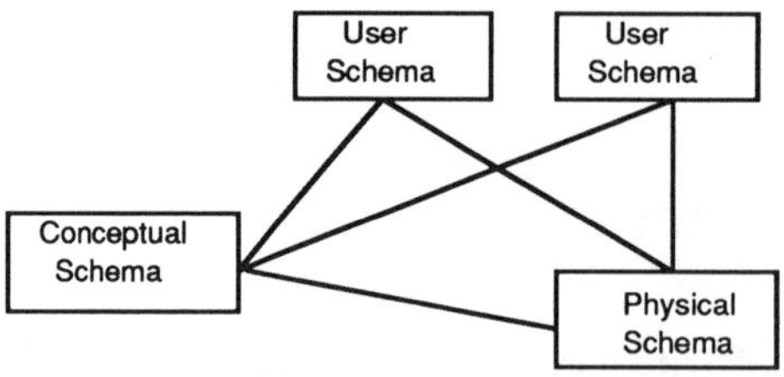

FIGURE 2 Three Schema Concept (Source: ANSI/X3/SPARC 1975)

— Storage and retrieval of data; selection of data using different fields.
— Standardized access to data and separation of the data storage and retrieval functions from the programs using the data. This makes database and application programs independent of each other, so that changes in one do not necessarily lead to changes in the other.
— A standardized interface between database and application programs to isolate the application program from the details of the physical storage structure.
— Access to the data by several users at a time.
— Definition and automatic enforcement of consistency constraints for the data. Consistency constraints are rules which must hold for all data stored, and are an excellent technique to reduce the number of errors in a large data collection.

For a slightly different, more elaborate list see Codd (1982).

The conceptual schema (Figure 2) describes the data and their relations at a logical level. It explains the potential linkages between the data and shows the potential for application programs and query languages to use the data. The physical or implementation schema maps the data which are described in the conceptual view to the storage devices. Separating logical from physical design allows the organization to maintain the conceptual structure of the data for a long time and easily adapt their mapping to accommodate changing hardware.

The conceptual schema adopts a 'corporate' view and attempts to show the logical integration of data from different organizational units and one can use multiple 'user schemata' to describe a conceptual view to each group of users that is more appropriate for their usage of the data (Frank 1990*c*). Typically a user view does not contain all the data available, but only the parts relevant for the task. In addition, it may also rename or restructure data to allow people to use the data in a structure with which they are familiar. The database management system then automatically converts the data to this user's 'view'.

The data must be considered a 'corporate' resource, and the more widely they are used the better. It is a general observation that economical benefits and thus a cost advantage of a database system are only attained if multiple user groups can use the same data. The more users that can share in the fixed cost of data collection and data update, the lower the costs incurred and the more beneficial the results. It must be possible for many people in an organization to access the same data at the same time and it may even be necessary for different groups to have well defined permissions in order to update the data concurrently. We also observe that organizations acquire geographic data from other agencies (Frank 1990*a*). This may lead to technical difficulties which can be overcome using concepts of transaction management (Date 1986, Ullman 1988).

It is not necessary for an organization to embrace a concept of centralization in order to reap the benefits of the database idea. Databases can be built conceptually centralized but physically decentralized, allowing each organizational unit to keep the data it needs but aiding an organized sharing of data (Frank 1990*a*).

A database management system must include methods to guarantee long term usability of the data in order to protect the original investment of the data collection effort. This includes the logical consistency constraints that guard against certain types of erroneous data. Methods must be set up to ensure that data, once collected, cannot be lost, despite malfunctioning of the hardware, disk failures or loss of electric current. The database management system must also help the organization to define responsibility for data and ensure that only authorized users update data elements following proper rules. They typically include controls to prevent unauthorized access to data. Cartographic data may be less sensitive and some mapping agencies less concerned with these problems, but a good database management system provides all these services — for very low additional cost — and they can occasionally be very useful.

Databases and Information Systems

The terms 'database' and 'information system' are often used in similar contexts — is there a difference? Before we access this issue, it is helpful to look first at the difference between data and information.

By *data* we understand sets of symbols in a form that is accessible to computer-based data processing. Computers can use these data and process them according to programs. The data can completely change their form, appearance and size but remain data. *Information*, on the other hand, is 'an answer to a human's question', a number of data that are meaningful to a person at a particular time. The human mind interprets the data in relation to the world and draws some conclusions, decides on action or similar. It is always a good idea to recall that computers do not make decisions! As an aside, we propose to use the word '*document*' for a collection of data in a format not currently accessible to a computer — so a map collection, records from an exploration, etc, are 'documents'. They need human interpretation to be meaningful, whereas the data are formalized so that programs can process them.

A database is thus a collection of data with its management software. The data in the database are accessible by other programs, but not directly by end users. The clients of the database are the programmers and the application programs they write. The information system delivers information to users, and it must therefore include methods (programs) for the users to communicate their 'information needs' and then other methods to present the data that make up the answer in a format that is understandable to end users. Thus an information system is more encompassing than just the database (see Figure 1) and includes specialized applications, query languages (Frank 1982, Egenhofer and Frank 1988) and cartographic output routines.

Spatial Databases

Both GIS and cartographic databases deal with the management of *spatial data*, which we define as data that have a relation to a spatial location (Frank 1987)

Spatial data = Geometry + Attributes

Cartographic databases and GIS deal with spatial data similarly to systems for CAD/CAM, VLSI, etc (Brodie 1984, Plouffe et al. 1984, Stonebraker and Guttman 1984, Udagawa and Misoguchi 1984, Wilkins and Wiederhold 1984). It has been observed that managing spatial data poses some special problems and that DBMS designed for the management of administrative data are not well suited to manage these non-standard data (Härder and Reuter 1985). We first address the functionality necessary for spatial data handling, specifically location-based retrieval. Afterwards we discuss how we model geometry to represent spatial data adequately.

It has been found that these non-standard applications require a larger set of functions than the standard commercial applications. There are

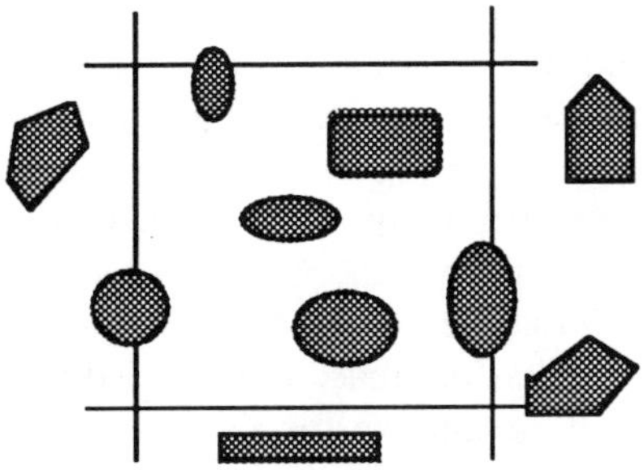

FIGURE 3 Query rectangle

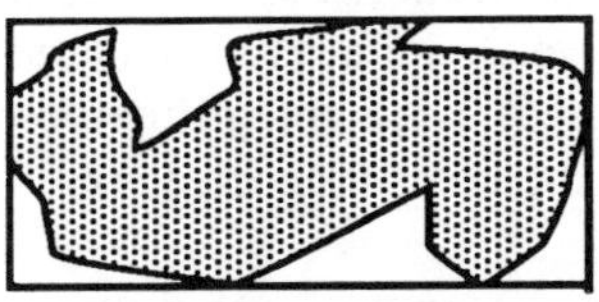

FIGURE 4 A minimal bounding rectangle

similarities among them in their requirements, but it is not clear if a single comprehensive DBMS for all these applications could or should be built. The cost of fulfilling the specific demands of spatial applications are high in terms of performance, hardware requirements, etc, and a non-spatial application should not have to bear this cost. Also, implementations need to be designed for the specific load, eg VLSI data have different characteristics from cartographic data. It has been proposed that a non-standard database should be built from a common kernel with extensions that could be freely combined to meet the demands of a specific application (Schek and Waterfeld 1986). Recent discussion has concentrated on the applicability of the object-oriented concept to databases (Dittrich and Dayal 1986, Dittrich 1988), to spatial databases and the specific problems of spatial access methods (Buchmann et al. 1990).

Efficient retrieval of spatial data based on spatial location is a special problem and some specific solutions have been proposed. The problem is generally stated as

'*Retrieve data for all objects within <rectangle>*'

where the rectangle (Figure 3) is given by the coordinates of two points.

In order for such queries to be answerable, the system must contain the minimal bounding rectangle (Figure 4) for each object (Frank 1981) or another method of expressing position and extension of an object in a generalized fashion, eg a bounding circle.

In order to be able to process such a query, which is technically speaking a two-dimensional range query, special data structures are necessary and a large number of methods have been developed (Frank 1983, Barrera and Frank 1989, Frank and Barrera 1990). For an overview see Samet (1989*a*) and for a recent conference on the subject consult Buchmann et al. (1990).

In systems that do not use such data structures, the database is divided into map sheets or facets and the user is aware of these. Special processes must be provided to join sheets and continue the objects from one to the other (Beard and Chrisman 1988). A *seamless* database not divided into sheets is necessary for a GIS and certainly attractive for a cartographic database (Chrisman 1990).

It is difficult to assess the differences in performance of these spatial data structures, as they depend on the specific properties of the data loaded and their performance seems to be relatively similar (Kriegel et al. 1990). The problem, not yet properly addressed, is to integrate these spatial data structures with the other parts of the comprehensive spatial DBMS.

The database discussion of the past few years has concentrated on the relational data model (Codd 1982, Ullman 1982) and many implementations are available today. Many commercial GISs allow one to store attribute data in relational databases and link them logically to the geometric or graphic data. The capability to connect to existing data sets stored elsewhere is important, but makes preserving consistency during update operations more difficult. It has been proposed to store all data, geometric and attribute, in a relational database (Waugh and Healey 1987, Abel 1988), but so far no convincing demonstration which includes a performance report of a realistically large data collection has been published. One major problem that remains is the large number of slow disk accesses that add up to unacceptable delays when retrieving the large data sets typically necessary to display a map.

It also appears difficult to fit the complex data structures of spatial reality into the restricted relational model (Roessel 1987). The relational data model was found to be quite restrictive, structuring all data in tables (relations) of a single type of tuple (record) with a fixed set of attribute values. Spatial data can be structured in this form, even if this is not necessarily the most natural description for spatial concepts and geometric data models (Frank 1990*b*, Goodchild 1990)(Figure 5).

The relational algebra on which relational databases are founded, process a relation at a time. This does not accord with the record at a time processing technique of the currently used programming languages. Programmers of spatial data handling routines, especially for geometric and topological operations cannot often exploit the power of relational processing and must resort to slow single record retrieval and processing. The difference in concepts between database and programming language further complicates the transfer of data values between the two.

Nodes (node-id, x, y)

Areas (area-id, property-value1, property-value2, ...)

Arcs (arc-id, id of start-node, id of end-node, id of left-area, id of right-area)

FIGURE 5 Relational data structure to implement the topological data model (Source: Herring 1990)

The object-oriented paradigm used in software engineering and programming language design (Meyer 1988) appears to be applicable to databases and promises to provide a more flexible method for structuring complex data (Egenhofer and Frank 1987). It is, similar to relational algebra, soundly based on a mathematical formalism, a property the earlier hierarchical and network data models missed (CODASYL 1971). An object-oriented data model should allow generalization hierarchies (probably several at a time, so called 'multiple inheritance') and methods of constructing complex objects from parts (Atkinson et al. 1989, Stonebraker et al. 1990). This appears to solve one of the major problems with the relational data model's breaking of conceptual objects into minute pieces (Nievergelt 1990) and makes the object-oriented data model most promising for spatial databases. Unfortunately, commercial quality object-oriented DBMS are just emerging and a number of research issues are still open (Kim 1990).

Cartographic Databases and GIS

A simple definition for a cartographic database is a database which contains cartographic data together with the management software necessary for its collection, update and output.

This begs the question what are *cartographic data*, a question which leads into the essence of designing a cartographic database. To begin with, it is sufficient to state that cartographic data are data from which maps and their graphics in one format or another can be produced or reproduced easily. A cartographic database is thus a database system — as defined above — that contains cartographic data, ie map data, and the procedures to display maps, either on a screen or on paper. These maps can then be used for different purposes.

In contrast, a *geographic information system* is an information system which can respond to a wider array of questions regarding some aspect of geography, of information about land, its use, ownership or population. The response can be in the form of a map, but can also be a report, a table or a textual answer.[2]

The major difference between a cartographic database and a GIS is in what is modelled. A GIS contains a 'model of reality' which is limited to some specific tasks and the major data needed for them. The cartographic database contains models of maps (which, in turn, are models of reality)(Chrisman 1983: 303, Frank 1984).It must be admitted that in practice there are no sharp distinctions and many systems which are advertised today as GISs are in effect more like cartographic databases, and some of the current cartographic databases do allow some extra flexibility in their use.

WHAT IS STORED IN A CARTOGRAPHIC DATABASE

In order to discuss the design of a cartographic database we have to understand what kinds of data should be stored. Database design should always proceed from a conceptual understanding of the data and from some rough assessment of the size of the data collection, to a discussion of the data structures to be used and then finally to the physical organization of the data.

What is stored in a database is described in the database schema. The database schema language provided with one or other of the generic DBMS is more or less capable of describing the data to be stored accurately and naturally (ie without any unnecessary artifacts). It is then that the descriptive powers of the data model and the database schema language become apparent.

A map can be viewed from different levels of sophistication, and depending on the viewpoint, different elements and structures become apparent (Chrisman 1983: 303–4). On a first level of differentiation, we can see a map as an image consisting of pixels, as a line drawing, or as a complex communication structure.

Maps Stored as Images

A simple view of cartographic data is to consider maps as images. The map surface is divided into regular cells (pixels, short for 'picture elements') and for each of them the colour and intensity is recorded. Such data can easily be collected by scanning existing maps and the maps later reproduced from the data. This is used extensively in navigation and military applications. The major drawback is the immense amount of data to be stored and processed even if compression methods are used. For example, a map sheet of 40cm × 60cm, scanned with a minimal resolution of 1/10 mm — about FAX quality — and four levels of intensity produces 6 Megabytes of raster data per colour and this at a very modest quality. Better quality resolution, eg 1200 dpi[1] ≅ 1/50 mm, and 256 levels of intensity results in 1·5 Gigabyte[2]. This is probably not the most desirable method of storing the thousands of large scale maps covering a large country.

More limiting than the immense amount of data to be handled are the limits on the operations which can be performed on such data: all the standard image processing operations, from averaging to edge enhancing are available, but there is only limited use for them in cartography. Images are well suited for 'electronic editing' at the pixel level, and this is used in high quality commercial map production. Hand-drawn manuscripts and other maps are scanned and then electronically combined and edited. If images from maps of different scales are stored, even a limited type of zoom is possible. Systems have been proposed that would support military command and control operations and would allow 'zoom' and 'pan' over an apparently seamless map (Tuori and Moon 1984). If the resolution is reduced to what is reproducible on a PC graphics screen (pixel *c* 1/10 mm), and some efforts in data compression, etc made, then a number of maps can be stored on a CD-ROM and with appropriate software they can be retrieved and used for a backdrop behind additional information managed in a separate format.

On a raster image, we cannot automatically find a location by name or condition on its location relative to other features. The system can be augmented with a gazetteer that lists named locations and their coordinates, and a translation algorithm from geographic coordinates to image coordinates. Then map displays can be retrieved by location name, but this is not an operation applied to the raster image properly, and uses additional data sets (Neal et al. 1989). On raster images, it is difficult to follow a line through intersections with other lines, and therefore all tasks of route planning, etc require human interaction to find the continuation of a line after a complex road intersection. It can also be difficult to find the area of certain complex figures when it is necessary to find the continuous boundary line. In general, raster images limit the analytical capabilities enormously. One could use the theory of symbolic projection (Chang 1990, Chang et al. 1990) on cartographic image data and create indexes in the form of strings. This would require segmenting the images into clusters of pixels that form one object, a notoriously difficult task in cartography, but it would then make additional retrieval logic available to the user (Chang et al. 1988).

Vectorized Data

Maps can be seen as a set of lines and points (disregarding colour filled areal features) and stored as such. Such vector data can be produced automatically from scanned maps by extracting the vectors from the raster. The necessary skeletonization process that provides vectorization proper has a tendency to deform junctions (Figure 6).

FIGURE 6 Skeletonization of a T–junction

Traditional rules for map symbolization do not always facilitate automatic digitalization. Separating lines based on colour and then assigning them a feature code is one of the few methods for automatically encoding features. Unfortunately this is hindered in cases where contour lines are printed in brown in general but blue for glaciers and lakes and black in rocky areas (eg some European topographic maps). Further, contour lines are often interrupted, not only to place index height figures, but also for buildings, etc. Filling these gaps automatically is difficult if not impossible.

At the conceptual level, the map is still an image to be reproduced by vectors, and vectorization is essentially a form of data compression but does not change applicable operations compared to image data. Names and labels are just sequences of vectors and not yet letters or symbols for which one can search.

Graphics Command Files

A map is drawn on a computer as a sequence of commands, which direct the plotting equipment to draw lines, characters, etc. A map can also be stored in this form, including commands to draw vectors in different line styles, text and symbols. Such a format is quite compact and at least for storage, quite efficient. It is the obvious choice for maps originally produced with an automated cartography system. As early work was primarily carried out at institutions that produced maps, this was the typical solution for a 'cartographic database' (Tomlinson 1988: 251), and some early work on data exchange standards was influenced by this. The data structures are simple linear sequences of coordinates, with some header records indicating the feature class (Canadian Council on Surveying and Mapping 1982, Zarzycki 1984). In principle this is still quite satisfactory for a system that primarily maintains maps, if one is aware of its limitations.

Data for such collections is also produced by digitizing existing maps using special programs which allow editing and cleaning of the data. The human digitizing operator interprets the drawing (the map document) to create a data set that has some higher level structure. For example, symbols and labels are encoded. Maps stored in this format can be updated with relative ease. The appearance of a symbol or the type face can be globally changed, which is not possible if we store the map as an image. The existing systems are primarily oriented towards map maintenance and do not provide extensive query facilities. It is not clear what would be the potential for access to such a database with an *ad hoc* query language.

Maps as Structures

A map is not just an image, at least not to a cartographically literate person: there are clearly distinct features which are considered separately within the spatial context. Most map users clearly see the advantage of using a road map over a remotely sensed image for planning a driving route; a road map contains the relevant facts (and few others) encoded in a format that is useful for route planning. In a remotely sensed image, the relevant information must first be extracted from a much larger, not yet filtered, collection.

One approach for structuring cartographic data is to consider cartography as a language with its own syntax and vocabulary (Dacey 1970, Robinson and Petchenik 1976). The formalization of this concept to a degree that is useful for cartographic data processing is slow to come (Youngmann 1977, Palmer and Frank 1988). It is clear that there is some regularity in the transformation of information into graphical symbols (Imhof 1972, Bertin 1983, Mackinlay 1986), which enables the map reader to correctly interpret the meaning of the symbols. However, we have not yet found a comprehensive set of rules that help to select these transformations (Buttenfield and McMaster 1991). If we consider the map as a collection of symbols which communicate some meaning and not as an image, we must be able to group the graphic elements into meaningful symbols and describe the relations between them — a very difficult problem (Schlichtmann 1984, Head 1991). The database schema then contains these symbols and their structure and not the graphical primitives with which they are drawn.

At this level, we may separate the map proper, the cartographic symbols that describe the world, and the context, that assigns meaning to the symbols and makes them understandable (Youngmann 1977). A structured description of the 'reference' part of a map is needed and some initial work was done by Ehrich et al. (1988) and Neumann (1988).

The 'information' part of the map can be structured initially in point features, line features, and areal features (Morrison 1988: 24). For each feature there is an indication of its location, either in map or real world coordinates, assuming that the transformation between the two is defined by a mathematical projection. This location is most often an approximation to the location of the real world entity on the surface of the earth, but it need not be. Maps also depict graphically, and thus spatially locate symbols for things that do not have a geometric location in reality, such as labels for highways, country names, and the symbols on thematic maps. Features are encoded using elaborate classification schemas, mostly based on cartographic (topographic) traditions (Canadian Council on Surveying and Mapping 1982), but other methods would be possible (Guttenberg 1981). There is sometimes a labelling text (eg the spot height, the name of the town) with a position for writing this text.

The US draft standard for digital cartographic data (Morrison 1988) lists geometrical structures for cartographic objects, defining terms like line segment, string, arc, etc for linear features and area, polygon, etc for areal features. These constructs should be sufficient to express the content of a map drawing, such that data can be exchanged from one system to another and be reproduced. They may also be a good base to start the design of the data structures of a cartographic database.

Unfortunately a single feature is not always mapped by a single map symbol nor does a graphical map symbol represent only one feature. One of the major problems of cartographic data structures is how to deal with the linkages between individual graphical map objects and the features. The structure of such a file approaches a state where it becomes a model of reality (ie a GIS): the coordinates are world coordinates, the feature classes are real object classes. There is a very subtle but crucial difference in understanding what is represented and for what purposes it exists. If one wants only to represent a map in a digital form, such that it can be edited, communicated and reproduced with computer systems, one is at one extreme. If one wants to construct a general purpose database from which to produce maps of varying content and scale and for different purposes one is in the middle, and the database for general purpose spatial analysis is at the other extreme. Again, one has to respect the special quality of a cartographic database and 'one cannot expect a database which was generalized using procedures for cartographic generalization to be a reliable source of predictable quality for analysis' (Brassel and Weibel 1988). It could also be added that one should not construct a database prepared for analytical tasks and expect a quality map. The acid test is often to determine if the database contains purely graphical data which has no counterpart in reality (eg the location of labels).

MAPPING PROCESS

In order to understand what is represented in a cartographic database, it may be useful to consider briefly the cartographic process. A cartographer goes through a number of steps to construct a new map and at each one

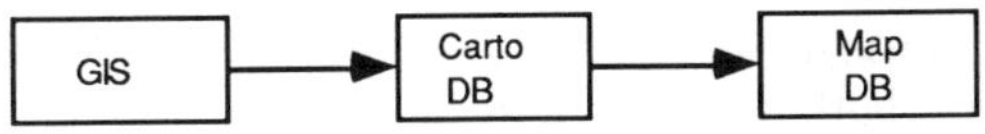

FIGURE 7 Deduce map database from GIS

applies different types of rules to different types of objects. These steps are roughly (Rase, Chapter 9):

— conceptualization of a map topic and content;
— collection of data;
— design of the map: the cartographer decides which part of his mental model of reality is to be shown on the new map and selects the topics to be rendered;
— design of the map style: the cartographer decides on the map symbols, colour schemata, etc;
— selection of the individual map features;
— rendering of map features.

The cartographer optimizes for various goals simultaneously, mostly to ensure maximum communication. This includes most aspects of the map generalization process. The cartographer often does not follow a strict linear sequence despite the fact that theoretical studies generally follow a linear sequence model. In principle one could deduce automatically the cartographic database from the GIS, as the manual cartographer establishes his data set from various sources (Figure 7).

Unfortunately, none of the steps of these processes are well understood and formally described. Despite some efforts, only very limited parts are automated. A few topics have been addressed, especially line generalization (Douglas and Peucker 1973, Beard 1988), label placements for point features, (Freeman and Ahn 1984) and generalization of built-up areas. One of the major problems is the lack of objective measure for the qualities for which one optimizes the map and which are affected by the generalization process (João 1991). For example, we lack a measure for the 'density' of a map and also do not have an operationally defined 'optimal density' which a programmed process could use. We also do not have appropriate methods to describe the 'scale' of a cartographic data set. There is agreement that the publication scale does not describe the essence, but it seems so far the only usable method.

As a theoretical experiment, let us assume for a moment that we have a fully programmed cartographic process, such that we can produce maps automatically from a GIS. It is not clear if such a fully automated process can ever be achieved, as criteria have not yet been set. What would be an acceptable level of performance? The problem appears to be 'AI complete', meaning that it requires 'understanding' (Brassel and Weibel 1988: 231) and if it could be solved, many other problems of artificial intelligence, eg understanding natural language text, might be solved. However, it appears entirely feasible that partial solutions can be found to solve a majority of the cases and produce results that are usable even if they are not of as high a quality as the best examples of current manual cartography. These processes could be made available to human cartographers and used under their control to support their work.

Continuing with the theroretical experiment, if the cartographic process was fully automatic we need only store a GIS as complete and as true to reality as possible from which we could deduce any map desired with no manual effort. There would, in principle, be no need for a cartographic database. If we do not have such an automated map production process, there is a need to store the intermediate results, because they are the product of human effort and should not be wasted. We assume here that human labour is valuable and needs to be economized and preserved, whereas computer processing power is essentially free and programmed steps can be repeated. Thus we see that a cartographic database can have any of the following contents:

— Cartographic model of the world, defining the real world entities which are of importance to map making (as opposed to real estate taxation for example). This essentially defines a map topic and a selection of object types to be included, but these data can still be used for analytical processes.
— Collection of data describing real world entities which should be mapped under certain circumstances, eg at a certain scale. This is the selection of individual features to be shown.
— Symbolized cartographic features, with labels positioned, etc, including many of the results of map generalization processes.
— Drawing instructions to produce a map.
— Image of the map.

First, one notices the similarity of this list with the results of the previous section. From this list it appears that the ultimate goal of a cartographic database is probably the 'cartographic model of the world' but the lack of a fully automated mapping process forces us to store intermediate steps. It seems as if the closer one comes to the final product, the less flexibility there is. This is largely correct, but the process is, as mentioned before, not linear. Decisions in each step have substantive effects on the total appearance of the map. For example, a simple change in the cut of the map sheet requires extensive adjustment of objects that were previously affected by the map boundary and must be amended (Figure 8).

The cartographic process is concerned with the management of 'map real estate'. A piece of paper (or screen) that is used to communicate one

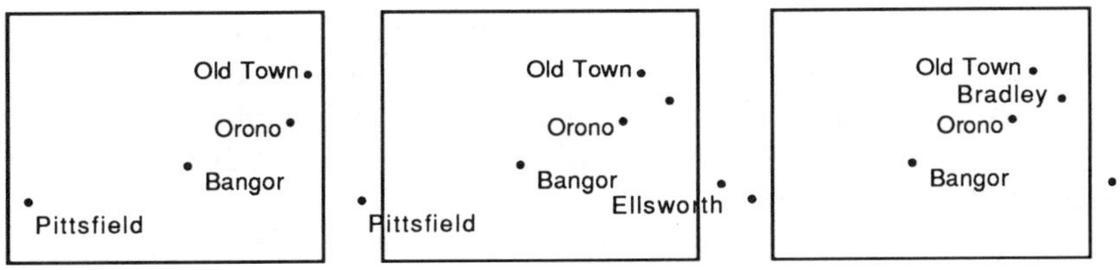

FIGURE 8 A change in map sheet size is affecting name placement

piece of information cannot be used for another one. Thus the spatial interaction of cartographic symbols on the map influences what must be selected and where the appropriate symbol must be placed. The cartographer tries to optimize this for maximum communication with the map user. Recording these elements of the graphical image makes the data collection less versatile and less useful in the production of other maps from the same data. Efforts underway to capture these rules (Buttenfield and McMaster 1991) may eventually help to overcome these limitations.

Many of the current efforts are directed not towards a fully automated mapping process, starting from a geographic information system or generic cartographic database, but towards a slightly different goal, namely the deduction of one map from an existing map data set, typically involving a change in scale (Beard 1987). This is certainly easier, if the given map is similar in scale, theme, etc to the desired one, but is still not quite feasible today (Dobson 1988).

LINKING A GIS AND A CARTOGRAPHIC DATABASE

Unless we can produce maps from a GIS without manual intervention and need maintain only one database (namely the GIS), we have to live with multiple databases that contain data about the same objects. The critical problem is not the original preparation of the derived data sets, but long-term maintenance. Without costly and error-prone maintenance, the data sets are out of date quickly and thus useless. For the following discussion, we will assume that the GIS is updated regularly in order to fulfil its analytical or administrative purposes. We will frame the problem in terms of automatic synchronization of the cartographic databases with the GIS updates. A change in the GIS database should be checked and, if necessary, the cartographic database automatically updated.

The cartographic database contains internal representations of specific maps that need to be produced and the GIS contains models of the same objects which can be used for other ends, for example spatial analysis. These two databases can, but need not, be separated and stored in different places. Nevertheless we can consider them as a logical unit and study their data models and schema. The methods for their synchronization are not substantially more difficult if the two databases are stored on different computers, the so-called 'distributed database' situation (Mohan 1984, Frank 1985).

In database jargon, one could consider the cartographic database as a 'materialized view' (Günther 1989), ie a specific way to present the contents of the database (thus in the database jargon a 'view') which is not produced ephemerally but is materialized (ie stored). This is a common method to avoid repeated costly processing (or in this case, processing which is not fully automated) and to save the human effort that went into construction of the view. The problem with materialized views is that they need to be updated with any change in the database (Blakeley et al. 1986, Hudson 1990).

The cartographic database contains all the data necessary to produce the map. It also contains linkage data that connect the cartographic features with the corresponding objects in the GIS (Broom 1989). These linkages are complex and need research attention, as they are not always just simple links between a GIS object like a building and the corresponding map feature. A forest feature on a map may be composed from a number of forest parcels, or an airport building may appear as a single feature in the cartographic database but as several objects in the GIS.

In its simplest form the linkage has no other objective than notification of change. Any change in the GIS is then translated into a warning flag associated with the relevant features in the cartographic database. If the cartographic database is used to produce a new version of the map the cartographer's attention is drawn to the flagged features and he individually and manually updates the cartographic features, which are then updated and stored in the cartographic database. The cartographer should have access to the updated data in the GIS, but there is no specific requirement or need for sophisticated generalization software; if it exists, it will increase the cartographer's productivity.

This notification of change should occur independently of the reason or the program used to change the GIS — any change should result in this notification. The functionality must thus be included in the DBMS software and not in the application programs of the GIS. In the database community this is called a 'trigger', which guards some database object and if a predetermined condition occurs (eg the object is changed) the trigger fires and a set of operations is executed, for example, the cartographic database is notified of the change. The database research community has investigated systems to achieve such effects in databases; results from the research in 'active databases' (Chakravarthy 1989) should be used for the cartographic database problem. The database researchers are usually concentrating on triggers in a single database, not loosely coupled databases like a GIS and a cartographic database. They have studied time-critical (less than one second) propagation of values (Dayal et al. 1988), but again this is not of concern in the cartographic database situation, where a notification within weeks is most likely sufficient. The requirements for the GIS and cartographic databases are thus slightly different. There is an immediate need to explore the exact demands and formalize specifications in order for the database research community to include our needs in their work (Smith and Frank 1990).

Practically, one will observe that many objects in the GIS and the corresponding features in the cartographic database contain exactly the same data. This depends on the model built in the GIS and the cartographic design. The greater the similarity between the purposes of the GIS and the map, the more correspondence between the data. The correspondence may become so large, that one forgets about the functional difference between a GIS and a cartographic database. This may be dangerous as there are most likely cartographic features which must be adapted for graphical representation; one is then tempted to tweak the GIS data to produce the correct cartography and lose the correct model of reality in the process.

A better solution would be to define a procedure that copies the data for the GIS in all cases where no change is necessary and uses data that are separately stored in the cartographic database in all other cases. The GIS data then serve as 'default' values which can be overridden by previously determined and stored cartographic data. Such a method can be extended to include some filtering of the data transformation or other programmed processing and only if this is not sufficient do specific values need to be stored and maintained. Current commercial DBMS, however, do not contain methods in their data models which facilitate this solution and extensions should be added. One can assume that such facilities would be useful for applications other than cartography and should therefore be constructed as a general purpose solution.

CARTOGRAPHIC DATABASES FOR MAP SERIES

Many of the map producing agencies and firms which are interested in a cartographic database produce not only a single map but multiple maps organized into series, which cover the same area and similar themes, eg for different uses or at different scales. Examples include the US Geological Survey that produces several topographical map series with different scales, and private cartographic companies which maintain data sets for road maps of different scale, map sizes, etc. The same cartographic features are then rendered in multiple maps in slightly different forms. It is a current challenge to develop database schemata that can represent this situation. In this case we deal with maintaining map data only, not linking map data with GIS data. A cartographic data model is used throughout and differences between the map series are limited. The majority of feature data appear the same on the different maps and considerable savings can be achieved by not copying these data. But again it is not the savings in storage space that are the most important, but the reduction in the expense of updating the map database which is the major benefit.

No implementation of such a multi-series cartographic database has been reported. Most organizations that maintain multiple map series are either not yet using computer aided systems and currently maintain independent graphical masters for each map in each series, or are just starting to use computer aided cartography. Even the few that have used computer cartography tools for a while maintain separate data files for each map sheet in each series, exactly paralleling the traditional organization (Dobson 1988). This is a common first step in automating any manual process, but should be overcome in the second phase, when one studies the special capabilities of a new technology to see how best to take advantage of them. In this case it is the possibility of maintaining in a coordinated manner multiple data sets with complex linkages which allows independent exploitation of each data set and permits transfer of changes from one to the other. Modern DBMS provide such facilities and they should be used to solve the difficult problem of maintaining map series.

As a practical example, one of the US commercial mapping companies maintains a large collection of road maps covering North America with many series of different scale, sheet cut, additional information, etc. They boast of including 19 536 changes since the previous year's edition in one single product (McNally 1990). Changes are mostly additions and deletions of road segments, changes in road numbers, etc, which for a large part affect multiple map sheets and must be manually applied to each — not forgetting the administrative effort of ensuring consistent updates of all maps. Replication of data in this example is not only between the different map series but also within one single series, where each map sheet covers a single state with some parts of the surrounding states; thus a change in a position close to a state border may affect one or more maps of adjoining states.

Often there is a difference in scale between map series. It is often said that digital cartographic data do not have a scale. Technically this is correct, as from one set of coordinate values (even if they are already scaled) a map in any other scale can be drawn, transformation of coordinate values from one map scale to another being simple. Photographic reduction or enlargement can only slightly change map scale and this is the case for computed coordinate transformations. Each map or data file representing a map has an intrinsic 'map scale' which was used to determine map content, symbolization, and level of generalization (Beard 1988). Currently we do not understand these limitations very well and cannot characterize a map data file other than by the scale for which it was constructed, a situation similar to the 'lineage' information used to characterize other data quality aspects of a data file, to which it is related (Chrisman 1985, Morrison 1988: 132).

We need to design a data structure for such a multi-series cartographic database and our first focus is on a data structure which supports multiple levels of detail. Computer scientists have studied a number of data structures that allow quick retrieval of the few salient aspects and increasing amounts of detail. Such structures are often built using the 'divide and conquer' approach. Examples are quadtrees and strip trees (Samet 1989 *a*, *b*). In such data structures, data are represented such that a rough approximation is obtained first and, with additional effort, is gradually refined. In a quadtree, areal data are represented with increasingly finer subdivision of squares. Rough approximations to an area can be constructed by considering only a few top levels of the tree and not accessing the detailed levels. This not only reduces detail in the perimeter but also excludes all areas smaller than a threshold. Combined with a method of smoothing the boundaries, results may be achieved that seem to be quite acceptable for certain applications (Bjørke and Aasgaard 1990).

The less well known strip trees (Ballard 1981) are based on a similar concept, but applied to line segments. A line is repeatedly approximated by two line segments, connecting the end points with the point furthest away from the straight connection between the end points. This can be applied recursively to the two newly constructed line segments. A line generalized to a certain level is retrieved by following the tree of approximations down to a certain level. This results in a line generalization somewhat similar to the well known algorithm described by Douglas and Peucker (1973). It seems easy to design a similar data structure comprising a collection of point symbols such that an increasing number are selected the deeper one searches.

These assume that

— the more refined representation is a superset of the less refined one; and
— one uses a regular method for subdivision and refinement.

Both these assumptions are not applicable to multi-scale cartographic data. In cartography, a map of larger scale is not just the data from the smaller scale map plus additional detail. We have to account for the case where a number of features of one kind are grouped together to form a generalized feature (eg a number of building symbols are replaced by a single town symbol), the generalized feature replacing the collection of

```
                                          — appears at 1:250,000
city,  symbol c1 (x, y),   label 'Atown'  pos (Δx, Δy)
                                          — move label at 1:200,000
                           label          pos (Δx, Δy)
                                          — shift symbol at 1:150,000
       symbol c2 (x,y)
                                          — add two labeled suburbs at 1:100,000
       symbol c3 (x,y),    label 'Asuburb'   pos (Δx, Δy)
       symbol c3 (x,y),    label 'Aport'  pos (Δx, Δy)
```

FIGURE 9 Changes in the data from scale to scale

details. Each feature has an associated map location where it is to be shown. Even for features that show on more than one map, their location may vary slightly from one scale to the next in order to accommodate other graphical map elements and avoid overcrowding. The data structure would have to use the values from the smaller scale map unless there are specific instructions stored with the larger scale level, each transition marked with the scale where the change or refinement occurs. For example one may see a sequence as in the following example (Figure 9).

In maintaining a multi-scale cartographic database one cannot identify a unique entry point for changes. Not all changes are entered in the database at the lowest scale and then percolate up. Changes are entered in the map sheet that is next to be produced and it seems desirable that no effort is spent at that time in updating the other sheets potentially affected — the cartographer needs to concentrate on the map he is working on to achieve a consistent product. Thus one should automatically detect which other maps could be affected and post a warning flag. If these sheets are to be re-edited, the cartographer sees all the updates that were posted on related sheets and he incorporates them as necessary in a manual editing step.

The methods sketched above appear to be practically useful, could be realized within a short period of time and promise considerable benefits. They do not require any methods of automatic cartographic generalization in order to work. Not only the reduction in storage space that could be achieved counts, but more importantly the simplified maintenance procedures. Ensuring that thousands of small changes are consistently applied on all maps is a major organizational problem.

PRACTITIONER'S SUMMARY

A cartographic database is a collection of data used for preparing maps together with the software to manage these data. Such databases are increasingly established by the map producing agencies and private companies and the data sets, together with the appropriate software, will be made available commercially to others that want to produce maps. Such databases will be organized in terms of computer graphics primitives (text, polyline, etc) and be ready to reproduce maps of acceptable quality.

A cartographic database is different from a GIS as a GIS is constructed for a multitude of applications and supports analytical procedures as well as some mapping; the GIS is intended to model a subset of reality geared to a specific set of tasks. The cartographic database is designed to manage map data.

A cartographic database can contain raster images of maps, produced by automated scanning of paper maps, vectorized map images, map drawing instructions or feature data. One arrives at similar differences if one considers the cartographic process and the intermediate products. Depending on the conceptual model used, different operations are available to the user. A difference easy to observe is the availability (or absence of) a query language. There the conceptual data structure is reflected in the kind of queries that can be executed. All users must always bear in mind the purpose for which a data collection was established and its limitations. Using a cartographic database for analytical work is likely to produce misleading results, and using a database established for analytical purposes to produce maps will result in a graphically inferior map.

Database management systems are the computer science tools for organizing a complex collection of data. They should be used to organize the storage of map data in a systematic fashion and help to protect the data for long time usage. The major investment is in the data collected, not in the hardware or software that manages it. Graphical map data have been used and reused for long periods of time and to achieve this situation in the electronic age we have to concentrate on the conceptual data structures in order to guarantee that the data collections can survive several changes in hardware and software. The database management tools help to achieve this.

One can logically link a GIS and a cartographic database for a region and establish connections between the objects in the GIS and the features in the cartographic database. If the GIS is updated the related features in the cartographic database can be flagged and cartographers warned about changes and will then manually effect the necessary cartographic changes. In such an arrangement one may also reduce the duplication of data between the GIS and the cartographic database and use the GIS for mapping purposes. This requires a 'default logic' and data structures which allow storage of specific map data that override or complement the GIS data, for example if features must be moved on the map to avoid graphically congested areas. The cartographic database, not the GIS, will also include positions for labels etc.

Somewhat similar organization could be used to manage the cartographic data for multiple map series of differing scale, content, etc, covering the same area. Currently such cartographic data sets are mostly stored independently in a file per map sheet. A database could add links between cartographic features representing the same objects in different maps and facilitate maintenance and guarantee the consistency of updates. From such a scheme methods of computer-assisted generalizations could be used to reduce gradually the data to be stored and hence manual effort. This seems immediately feasible, does not require any scientific breakthroughs and promises considerable benefits.

PROJECTIVE SUMMARY

A database must respect the inherent structure of the application area with which it deals. Thus the most important aspect of a cartographic database is to design its structure based on an understanding of the cartographic production and communication process. We are severely lacking in this. Currently cartographic databases are mostly established in terms of computer graphics output primitives. A more appropriate structure could start with understanding maps as a language and using a cartographic grammar as a means of structuring the database (Brassel and Weibel 1988: 235). The structure revealed by a grammar is, in principle, translatable into a database structure.

Despite a copious literature on map generalization (for an overview see Beard (1988) and Brassel and Weibel (1988)) very little is sufficiently well known that appropriate computer programs can be written. It is quantitative descriptions of map data and quantitative characteristics of different types of map data or cartographic processes that are mostly lacking. A notable exception is Buttenfield's effort to characterize cartographic line data in order to guide the appropriate line generalization algorithm and its parameters (Buttenfield 1989). Except for precision of point and contour lines, we currently lack methods to describe quantitatively the quality of cartographic data, etc, and abuse 'scale' and expectations of the current mapping practice to characterize cartographic data sets. Progress in methods of describing map data are urgently needed and will benefit generalization studies, as we gain quantitative goals for the optimizations involved.

Another line of investigation is to analyse the architecture of a cartographic production software and identify what steps are necessary in the map rendering process, how they are structured and what data are necessary for each. Using principles of software engineering to organize processing in layers with similar functions and achieve minimal interaction between modules will reveal the structure of the data needed and how they are interrelated.

An immediate line of research and development should be to structure the multi-representational cartographic databases that are required. This is not only a matter of practical concern but yet another approach to analysing cartographic data structure, this time starting from real cartographic data and their analysis. The goal would be to design a data structure that can manage the data sets necessary to produce the map sheets of several map series, which cover the same area, but at different scale or with differences in map styles. Two types of benefits can be reaped; namely the reduction of data storage (due to reduction in redundancy) and, more importantly, simplifications in the map update process.

ACKNOWLEDGEMENTS

Funding from NSF for the NCGIA under grant SES 88–10917 is gratefully acknowledged.

REFERENCES

Abel D (1988) Relational data management facilities for spatial information systems. In: Marble D (ed.) *Third International Symposium on Spatial Data Handling*. Sydney, Australia

ANSI/X3/SPARC (1975) Study Group on Database Management Systems, Interim Report 75–02–08. *SIGMOD FDT* 7 (2)

Atkinson M, Bancilhon F, DeWitt D, Dittrich K, Maier D, Zdonik S (1989) The object-oriented database system manifesto. In: *First International Conference on Deductive and Object-oriented Databases* Kyoto, Dec. 1989, IEEE.

Ballard D H (1981) Strip trees: A hierarchical representation for curves. *Communications of the ACM* **24** (5): 310–21

Barrera R, Frank A U (1989) Analysis and comparison of the performance of the fieldtree. Report. Orono, ME, Department of Surveying Engineering, University of Maine

Beard K (1987) How to survive a single detailed database. In: Chrisman N R (ed.) *AUTO-CARTO* 8, *Eighth International Symposium on Computer-Assisted Cartography* 1987. Baltimore, MD

Beard K (1988) *Multiple representations from a detailed database: A scheme for automated generalization*, PhD Thesis, University of Wisconsin – Madison

Beard M K, Chrisman N R (1988) Zipper: A localized approach to edgematching. *The American Cartographer* **15**: 163

Bertin J (1983) *Semiology of Graphics*. Madison, WI, The University of Wisconsin Press

Bjørke J T, Aasgaard R (1990) Cartographic zoom. In: Brassel K and Kishimoto H (ed.) *4th International Symposium on Spatial Data Handling* July 23–27 1990. Zurich, IGU: 859–68

Blakeley J A, Larson P-A, Tompa F W (1986) Efficiently updating materialized views. In: Zaniolo C (ed.) *SIGMOD* '86 May 28–30 1986. Washington DC, ACM SIGMOD Record: 61–71

Brassel K E, Weibel R (1988) A review and conceptual framework of automated map generalization. *International Journal of Geographic Information Systems* **2** (3): 229–44

Brodie M L (1984) CAD/CAM database management. *IEEE Database Engineering* **7** (2): 12–20

Broom F (1989) Multiple representations — U. S. Bureau of the Census position. In: Buttenfield B P and DeLotto J S (ed.) *Multiple Representations*: *Initiative* 3 *Specialist Meeting Report*. Santa Barbara, CA, NCGIA: 72–3.

Buchmann A, Günther O, Smith T, Wang Y-F (1990) *Design and Implementation of Large Spatial Databases. Proceedings of the First Symposium SSD* '89, Santa Barbara CA, July 1989. New York, Springer Verlag (*Lecture Notes in Computer Science* 409)

Buttenfield B, McMaster R (1991) *Rule based cartographic generalization* (in preparation)

Buttenfield B P (1989) Scale-dependence and self-similarity of cartographic lines. *Cartographica* **26** (1): 79–100

Canadian Council on Surveying and Mapping (1982) *National Standards for the Exchange of Digital Topographic Data — Draft Report*. Ottowa, Energy, Mines and Resources Canada

Carroll L (1960) *The Annotated Alice. Alice's Adventure in Wonderland and Through the Looking Glass*. Hardmandsworth, UK, Penguin Books Ltd

Chakravarthy S (1989) Rule management and evaluation: An active DBMS perspective. *SIGMOD Record* **18** (3): 20–8

Chang S-K (1990) Visual reasoning for information retrieval from very large databases. *Journal of Visual Languages and Computing* **1** (2): 41–58

Chang S-K, Jungert E, Li Y (1990) The design of pictorial databases based upon the theory of symbolic projection. In: Buchmann A, Günther O, Smith T R and Wang Y-F (ed.) *Design and Implementation of Large Spatial Databases*. New York, Springer Verlag: 303–24 (*Lecture Notes in Computer Science* 409)

Chang S K, Yan C W, Arndt T, Dimitroff D (1988) An intelligent image database system. *IEEE Transactions on Software Engineering* (5): 681–8

Chrisman N (1985) An interim proposed standard for digital cartographic data quality: Supporting documentation. In: Moellering H (ed.) *Digital Cartographic Data Standards*: *An Interim Proposed Standard*. Columbus, OH, National Committee for Digtial Cartographic Data Standards

Chrisman N R (1983) The role of quality information in the long-term functioning of a geographic information system. In: Wellar B S (ed.) *Auto-Carto* 6 Oct. 16–21 1983. Ottawa: 303–312

Chrisman N R (1990) Deficiencies of sheets and tiles: Building sheetless databases. *International Journal of Geographical Information Systems* **4** (2): 157–67

CODASYL (1971) *Data Base Task Group Report*. New York, ACM

Codd E F (1982) Relational data base: A practical foundation for productivity. **25** (2): 109–17

Dacey M F (1970) Linguistic aspects of maps and geographic information. *Ontario Geographer* **5**: 71–80

Date C J (1986) *An Introduction to Database Systems*. Reading, MA, Addison-Wesley Publishing Company

Dayal U, Blaustein B, Buchmann A, Chakravarthy U, Hsu M, Ledin R, McCarthy D, Rosenthal A, Serin S, Caery M J, Livny M, Jauhavi R (1988) The HiPAC project: Combining active databases and timing constrains. *SIGMOD Record* **17** (1): 51–70

Dittrich K R (1988) *Advances in Object-Oriented Database Systems — 2nd International Workshop on Object-Oriented Database Systems*, Bad Münster am Stein-Ebernburg, FRG, Berlin, New York, Springer Verlag (*Lecture Notes in Computer Science* 384)

Dittrich K R and Dayal U (1986) *Proceedings International Workshop in Object-Oriented Database Systems*. Washington, DC, IEEE Computer Society Press

Dobson M W (1988) Digital cartography in the world of commercial publishing. In: Marble D (ed.) *Third International Symposium on Spatial Data Handling* 1988. Sydney, Australia, International Geographical Union: 1–8

Douglas D H, Peucker T K (1973) Algorithms for the reduction of the number of points required to represent a digitized line or its caricature. *Canadian Cartographer* **10** (2): 112–22

Egenhofer M J, Frank A U (1987) Object-oriented databases: Database requirements for GIS. In: Aangeenbrug R T and Schiffman Y M (ed.) *International Geographic Information Systems (IGIS) Symposium: The Research Agenda* Nov. 15–18 1987. Crystal City, VA, NASA: 189–212

Egenhofer M J, Frank A U (1988) Towards a spatial query language: User interface considerations. In: DeWitt D and Bancilhon F (ed.) *14th International Conference on Very Large Data Bases* Aug. 1988. Los Angeles, CA: 124–33

Egenhofer M J, Frank A U (1990) Prospective views of GIS technologies and applications. In: *Symposium Brasileiro de Geoprocessamento* 1990. São Paulo, Brazil, Escola Politécnica da Universidade de São Paulo: 95–102

Ehrich H, Lohmann F, Neumann K, Ramm I (1988) A data base language for scientific map data. *Geologisches Jahrbuch* A (104): 139–52

Frank A U (1980) Landinformationssysteme — Ein Versuch zu einer Abgrenzung (Land Information Systems — An attempt of a demarcation). *Nachrichten aus dem Karten- und Vermessungswesen*. Frankfurt a.M., FRG, Institut für angewandte Geodäsie

Frank A U (1981) Applications of DBMS to Land Information Systems. In: Zaniolo C and Delobel C (ed.) *Seventh International Conference on Very Large Data Bases VLDB* Sept. 1981. Cannes, France: 448–53

Frank A U (1982) MAPQUERY: Database query language for retrieval of geometric data and its graphical representation. *ACM SIG Computer Graphics* **16** (3): 199–207

Frank A U (1983) *Probleme der Realisierung von Landinformationssystemen*; 2. *Teil*: (*Problems in the realization of Land Information Systems*; *Second part*:) *Storage methods for space related data*: *The FIELD TREE*[reprint of the paper in

Barr M (ed.) *Spatial Algorithms for Processing Land Data Symposium* 1983.] Bericht 71. Zurich, Switzerland: Swiss Federal Institute of Technology, Institute for Geodesy and Photogrammetry.
Frank A U (1984) Computer assisted cartography: Graphics or geometry? *Journal of Surveying Engineering. American Society of Civil Engineers* **110** (2): 159
Frank A U (1985) Distributed databases for surveying. *Journal of Surveying Engineering, American Society of Civil Engineers* **111** (1): 79–88
Frank A U (1987) Overlay processing in spatial information systems. In: Chrisman N R (ed.) *AUTO-CARTO* 8, *Eighth International Symposium on Computer-Assisted Cartography* March 29 – April 3 1987. Baltimore, MD: 16–31
Frank A U (1988) Requirements for a Database Management System for a GIS. *Photogrammetric Engineering & Remote Sensing* **54** (11): 1557–64
Frank A U (1990*a*) Acquiring a digital base map — A theoretical investigation into a form of sharing data. *URISA Journal* (in press)
Frank A U (1990*b*) Spatial concepts, geometric data models and data structures. In: *GIS Design Models and Functionality* Mar 21–22 1990. Leicester, UK, Midlands Regional Research Laboratory, University of Leicester
Frank A U (1990*c*) Telecommunication and GIS — Opportunities and challenges. In: Cavill M (ed.) *Networking Spatial Information Systems* May 30–31 1990. Melbourne, Telecom Australia: AF – 1–AF – 9
Frank A U, Barrera R (1990) The Fieldtree: A data structure for Geographic Information Systems. In: Buchmann A, Günther O, Smith T R and Wang Y-F (ed.) *Design and Implementation of Large Spatial Databases*. New York, Springer-Verlag: 29–44 (*Lecture Notes in Computer Science* 409)
Freeman H, Ahn J (1984) Autonap — An expert system for Automatic Map Name Placement. In: *International Symposium on Spatial Data Handling* 1984. Zurich, Switzerland: 544–69
Goodchild M F (1990) A geographical perspective on spatial data models. In: Maguire D (ed.) *GIS Design Models and Functionality* 21–22 March 1990. Leicester, UK, Midlands Regional Research Laboratory, University of Leicester
Günther O (1989) Database support for multiple representation. In: Buttenfield B P and DeLotto J S (ed.) *Multiple Representations: Initiative* 3 *Specialist Meeting Report*. Santa Barbara, CA, NCGIA: 50–2
Guttenberg A (1981) Uniformity and flexibility in the classification of topographic data. In: *FIG* 1981. Montreux, Switzerland
Härder T, Reuter A (1985) Architecture of database systems for non-standard applications (in German). In: Blaser A and Pistor P (ed.) *Database Systems in Office, Engineering, and Science*. New York, Springer Verlag
Head C G (1991) Mapping as language or semiotic system: Review and comment. In: Mark D M and Frank A U (ed.) *Cognitive and Linguistic Aspects of Geographic Space*. Dordrecht, Kluwer Academic Publishers
Herring J R (1990) TIGRIS: A data model for an Object Oriented Geographic Information System. In: Maguire D (ed.) *GIS Design Models and Functionality* 21–22 March 1990. Leicester, UK, Midlands Regional Research Laboratory, University of Leicester
Hudson D L (1990) Autonomous view-stages: Materialized support for view update propagation. PhD Thesis, University of Maine
Imhof E (1972) *Thematische Kartographie*. Berlin

João E M (1991) What expert systems don't know: The role of the user in GIS generalization. In: Mark D M and Frank A U (ed.) *Cognitive and Linguistic Aspects of Geographic Space*. Dordrecht, Kluver

Kim W (1990) Object-oriented databases: Definitions and research directions. *IEEE Transactions on Knowledge and Data Engineering* **2** (3): 327–41.

Kriegel H-P, Schiwietz M, Schneider R, Seeger B (1990) Performance comparison of point and spatial access methods. In: Buchmann A, Günther O, Smith T R and Wang Y-F (ed.) *Design and Implementation of Large Spatial Databases*. New York, Springer-Verlag: 89–114 (*Lecture Notes in Computer Science* 409)

Mackinlay J (1986) Automating the design of graphical presentations of relational information. *ACM Transactions on Graphics* **5** (2): 110–41

McNally R (1990) *Road Atlas*. Chicago, Rand McNally Corp

Meyer B (1988) *Object-Oriented Software Construction*. New York, Prentice Hall

Mohan C (1984) *Tutorial: Recent Advances in distributed Data Base Management*, Annotated Bibliography. Silver Spring, MD, IEEE Computer Society Press

Morrison J (1988) The proposed standard for digital cartographic data. *The American Cartographer* **15** (1): 9–140

Neal J G, Thielman C Y, Dobes Z, Haller S M, Glanowski S, Shapiro S C (1989) CUBRICON: A multi-modal user interface. In: *GIS/LIS* 89 1989. Orlando, FL

Neumann K-H (1988) *A Geoscientific Database Language for User Defined Geometric Data Types* (in German). PhD Thesis, University Braunschweig

Nievergelt J (1990) 7 ± 2 Criteria for assessing and comparing spatial data structures. In: Buchmann A, Günther O, Smith T R and Wang Y-F (ed.) *Design and Implementation of Large Spatial Databases*. New York, Springer-Verlag: 3–28 (*Lecture Notes in Computer Science* 409)

Palmer B, Frank A (1988) Spatial languages. In: Marble D (ed.) *Third International Symposium on Spatial Data Handling* 1988. Sydney, Australia: 201–10

Plouffe W, Kim W, Lorie R, McNabb D (1984) A database system for engineering design. *IEEE Database Engineering* **7** (2): 48–55

Robinson A, Petchenik B (1976) *The Nature of Maps*. Chicago, IL, University of Chicago Press

Roessel J V (1987) Design of a spatial data structure using the relational normal forms. *International Journal of Geographical Information Systems* **1** (1): 33–50

Samet H (1989*a*) *Applications of Spatial Data Structures. Computer Graphics, Image Processing and GIS*. Reading, MA, Addison-Wesley Publishing Co

Samet H (1989*b*) *The Design and Analysis of Spatial Data Structures*. Reading, MA, Addison-Wesley Publishing Co

Schek H, Waterfeld W (1986) A database kernel system for geoscientific applications. In: Marble D (ed.) *Second International Symposium on Spatial Data Handling*. Seattle, WA, 273–88

Schlichtmann H (1984) Characteristic traits of the semiotic system 'map symbolism'. *The Cartographic Journal* **22**: 23–30

Smith T R, Frank A U (1990) Report on workshop on very large spatial databases, *Journal of Visual Languages* **1** (3): 291–309

Stonebraker M, Guttman A (1984) Using a relational database management system for computer aided design data — An update. *IEEE Database Engineering* **7** (2): 56–60

Stonebraker M, Rowe L A, Lindsay B, Gray J, Carey M, Brodie M, Bernstein P,

Beech D (1990) *Third-generation Data Base System Manifesto*. Memorandum No. UCB/ERL M90/28. UC Berkeley, Electronics Research Lab

Tomlinson R F (1988) The impact of the transition from analogue to digital cartographic representation. *The American Cartographer* **15** (3): 249–61

Traub J F (1989) *Scaling Up: A Research Agenda for Software Engineering*. Washington DC, National Academic Press

Tuori M, Moon G C (1984) A topographic map conceptual data model. In: 1*st International Symposium on Spatial Data Handling* Aug 20–24 1984. Zurich, Switzerland, 28–37

Udagawa Y, Misoguchi T (1984) An extended relational database system for engineering data management. *IEEE Database Engineering* **7** (2): 67–75

Ullman J D (1982) *Principles of Database Systems*, 2nd Edition, Computer Software Engineering Series. Rockville, MD, Computer Science Press

Ullman J D (1988) *Principles of Database and Knowledgebase Systems Vol* 1, Principles of Computer Science Series. Rockville, MD, Computer Science Press

Waugh T, Healey R (1987) The GEOVIEW design: A relational database approach to geographical data handling. *International Journal of Geographical Information Systems* **1** (2): 101–18

Wilkins M, Wiederhold G (1984) Relational and entity–relationship model databases and VLSI design. *IEEE Database Engineering* **7** (2): 61–6

Youngmann C (1977) A linguistic approach to map description. In: Dutton G (ed.) *First International Advanced Study Symposium on Topological Data Structures for Geographic Information Systems* Oct. 16–21 1977. Cambridge, MA, Laboratory for Computer Graphics and Spatial Analysis, Harvard University: 1–17

Zarzycki J M (1984) Standards for digital topographic data — The Canadian experience. *Surveying and Mapping* **44** (1): 29–37

Zehnder C A (1981) *Informationssysteme und Datenbanken*. Zurich, Switzerland, Verlag der Fachvereine

Notes

[1] 1 Megabyte (MB) = 10^6 bytes (1 million characters), 1 Gigabyte (GB) = 10^9 bytes.

[2] For purposes of this chapter, we consider LIS (Land Information Systems)[FIG, 1981 #679, Resolution 302] AM/FM (Automated Mapping/Facilities Management) and similar systems as subclasses to GIS

Chapter 3

DEVELOPMENT OF CARTOGRAPHIC DATABASES

François Salgé
Institut Géographique National, B.P. 68, 94160 Saint-Mandé, France

INTRODUCTION

A cartographic or a geographic database, that is the first decision to be made when designing a database. Cartographic information may be seen as a description of the real world with the aim of producing maps (surface structure). Thus, the information attached to a cartographic object refers to its graphic description, ie elements of a legend, width, shape or colour; of a line or pattern; of an area or symbol; or of a point. Geographic information is a semantic description of the real world (deep structure) where geographic objects are given digital representations in such a way that the computer knows what is a road, a bridge, a river, with their associated attributes and links. (Salgé 1989, ATKIS 1990, Moellering 1990, Morehouse 1990). In the context of this paper the acronym GDB stands for a Geographic Database or a Cartographic Database with special comments whenever the latter have to be considered. They are generally accepted as nationwide databases.

Developing a GDB implies first defining its scope. Only a specific view of the real world has to be 'mapped' into the GDB, ie a definition of the geographic objects and their related attributes and the links between them. A phase of structure definition is then undertaken using a conceptual data model (CDM) leading to the conceptual data schema (CDS) which reflects the structure of the information. (Salgé 1989, Guptill 1990). Sources and processes, the way of storing and retrieving the data, and the way of using them also have to be specified.

From the systemic point of view, a GDB is composed of four subsystems. Those systems, which are well-known to the database community, are : capturing, handling, processing and reporting (Gardarin 1982, Pressense and Salgé 1987). However for digital geographic information, each of those subsystems can be considered as a complete system.

The data capturing (or data collecting) system requires the data to be incorporated into the database, and elements for updating existing information. Generally speaking, it has to turn analogue information into digits. The data handling system has to store, manage and retrieve the information. The data processing system is in charge of the analysis, crossing, and calculations using the data in order to make deductions from the stored and retrieved information. Data reporting or data exploitation tries to turn analytical results into material understandable by humans via readable reports, maps or drawings, (the best user interface for digital geographic information being a graphic output).

As previously suggested, a pre-requisite for GDB development is data content definition and a post-requisite is the orgware and the operationality of the system.(Aangeenbrugg and Schiffman 1987, Rhind 1988). As a GDB (from the decision to define one to its extinction) may last several decades, the human organization and the standard process definition are key conditions for a successful and useful system. Costs may be estimated as 20 % for software and hardware (10 % each), 40 % for data and the remaining 40 % for orgware (MARI 1990).

DATA CONTENT DEFINITION

Recently, existing GDB evolution has shown an important change to accommodate data requirements: from a producer defined set of data to a user defined set of data. During the same period, the tasks for which the data were being used have become increasingly diverse, placing information demands on the data that were not foreseen in their initial design. (Guptill 1990). This situation led producer agencies to define new feature-based spatial data model designs. This trend signals an evolution in the design of spatial databases and GIS.

Issues related to flexibility, standardization and compatibility require changes in data content, geographic coverage, accuracy, and consistency among data categories. All of them have an impact on the design of the data model. Five levels can serve as steps in refining a design to meet a set of requirements: (Moellering 1984, Guptill 1990, Salgé 1990). (Figure 1):

(1) *Reality*, which is the total phenomenon as it actually exists.
(2) *Data reality*, which is an abstraction of the reality that includes only those entities thought to be relevant to anticipated needs (scope of the data).

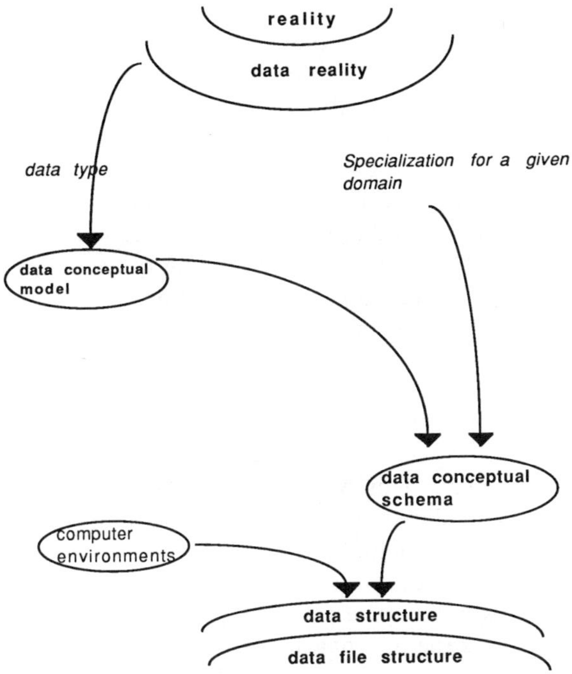

FIGURE 1 Five steps in refining a design to meet a set of requirements

(3) *Data conceptual schema* (*DCS*), which specifies the sets of components and relationships among the components pertaining to the specific phenomena defined by the data reality (system and structure independent); this DCS lies on a *data conceptual model* (*DCM*), which is a modelling framework enabling the concepts to be organized regardless of the data reality to be modelled.
(4) *Data structure*, which specifies the logical organization of the components of a data conceptual schema and the manner in which relationships among components are to be explicitly defined within computing system environments.
(5) *Data-file structure*, which is the physical organization of the data.

Reality and data reality are not of research concern. They impose themselves as sectors where digital geographic information may or may not apply. Only a better understanding of the scope of the domain and how the phenomena imply a geographic interpretation are required.

The data conceptual model is a concern for further research. The models as they are known from the database environment (relational, object-oriented) show how inefficient they are for handling geographic information (Bernstein et al. 1988, David 1990). Concepts, such as topology, objects, classes, attributes (and values), and links, are efficient for modelling all kinds of known geographic data but they also have some well-known limitations. These include methods (operators, integrity constraints, rules) which cannot be attached to objects (Morehouse 1990) and still be considered as methods (in the sense of object-oriented methods), quality which cannot be attached to the entities as quality information (the model only knows them as attributes), and a pure three dimensional model has to be found as existing ones are $2\frac{1}{2}$ models, etc.

Scaleless and seamless databases, taking history into account, are possible in the future (Guptill 1989)), and will lead, among other things, to a DCM able to cope with multiple geometric representations of objects.

The definition of a data conceptual schema implies the identification of the data reality as perceived for the given domain. This identification corresponds to the unambiguous definition of the classes of objects, their attributes and the links between them, with, for each of them, the association of information such as exact descriptions (exhaustive definition), quality requisites, etc. This will lead to the writing of specifications of the data conceptual model which is the bible of the GDB (Salgé and Sclafer 1988). Tools to help the database designer which simplify his work and lead to better solutions are also a concern of future research. (Bouzeghoub 1986). Rules may be defined and the use of expert systems should be considered. As the international context leads to data exchange between nations, an international data conceptual schema (multilingual) will arise leading to standards which make it possible to map every nationwide database in that internationally agreed schema. (Salgé 1990, Smith 1990, Sowton 1990).

As the techniques employed for DCM and DCS design become increasingly complex, the skills of the staff will have to be improved, and systems and methods which hide this complexity will be welcome. (Chorley 1988, Burns and Henderson 1989, Goodchild 1990)

DATA CAPTURE

As previously stated any database relies on data which have been digitized. The amount of work involved in data capture is a major task. Updating a database is the concern of the data producer and the user, as the reality, evolves due to human activities. Data capture and data updating (ie digitizing of updated data) implies tremendous manual work as the main concern is to turn analogue information into digital data. Existing databases have been loaded with data from maps (by digitization or scanning), from air survey (by stereo-plotting), remote-sensing imagery (by image processing, pattern recognition, or classification), field completion (digital measurement) and from other databases (incorporation of existing databases).

All those procedures rely on processes where human activities are predominant. Basically, they are all based on six steps:

(1) preparation of data and extra information gathering
(2) digitizing
(3) data checking
(4) data correction (steps 3 and 4 involve iterations)
(5) final data control
(6) data integration into the data base

The challenge for the future is to make each of those steps as automatic as possible.

Steps 2 and 4 are closely related as they imply techniques which are similar.

Steps 3 and 5 are also closely related as the former is based on a Quality Assurance plan (a kind of checklist to be followed as the digitizing process is in progress) and the latter is based on a Quality Control plan (a kind of final checklist to be applied, a sort of final quality evaluation of the data set : correct or repeat the procedure).

From those six steps, four research themes can be defined. As operators are not necessarily able to entirely understand the complete and complex structure of the data within the DBMS, there is a challenge to define capture-oriented data structures which are human interfaces to the data structure, a 'natural' and obvious interface to it. Those capture-oriented schema are necessary for obtaining a real operational process and a second challenge is to make it robust against operator mistakes. They are, in fact, procedures which imply sets of rules where a knowledge-based approach might be useful. As the key point of the setting-up of such a process is the efficiency from the operator points of view, efficient knowledge-based systems have to be developed and that can be seen as a third challenge. As the six previously defined steps are functionally very close to GISs, one can imagine using an existing GIS for that data capture process. (Goodchild 1990, Salgé, 1990). The fourth challenge is that as existing GISs cannot really be operated by non-expert operators, even though some fourth generation language is associated with them, an efficient system for semantic information capture is needed.

Step 1 can hardly be automated as the data gathering and preparation seems to rely to some extent on existing drawings. As our society is turning steadily towards a completely computerized environment, databases will exist which will include elements useful for the updating of a GDB, eg new highways or interchanges are designed using CAD/CAM systems, and the challenge here is to insert those data into the GDB.

Steps 2 and 4 are gradually being turned into more automatic processes. The challenge here is to define an automatic process for extracting information from air surveys, remotely-sensed data, and rasterized maps.

(NCGIA 1989, Ilg 1990). The techniques to be used are specific image analysis methods, those found in the pattern recognition literature are not as efficient as needed. The problem to be solved is one of 'geographic object retrieval from digital images of the real world'. Can we dream of fully automatic object extraction from digital air surveys by AD2010 ? An alternative would be to use the GPS in cars which would make it possible to 'drive' the updating of a GDB by an in-car system. Efficient methods for capturing update data also need to be found: via digital air surveys, and GPS devices, and by re-use of data from 'Design' Systems. Fuzzy objects (those whose boundaries are not well defined) still cause trouble as there are few ways of capturing and defining them.

Steps 3 and 5 relate to Quality Assurance and Quality Control plans. During the entire digitizing process, from input data to the sending of information to the DataBase Management System, a step-by-step quality procedure has to be set up and final quality measurements have to be established. Even though those procedures have in the past been implemented with some success, there remains the digital definition of quality (Bickmore and McCullagh 1989, Chrisman 1989, Drummond 1988). What are the quality criteria to be measured and how does one measure and express them? (Goodchild, Chapter 6).

DATA HANDLING AND DATA MANAGEMENT

Handling data in the context of a database management system, whatever it should be, is based on three kinds of schemas.(Gardarin 1982).

(1) Conceptual schema express the Data Conceptual Schema of the GDB in an understandable way for the DBMS. They translate the DCS using the tools and following the model which underlies the system. Present DBMSs show their inefficiency for handling the fact that a GDB deals with spatial data. At least three techniques can be followed: extended relational database; object-oriented database, (Kemp 1990) 'DBMS generator'. (David 1990). For each of those techniques one can try to evaluate their ability for handling spatial data (which data type, which algorithm) and their efficiency for storage and retrieval. Semantic, topologic and metric data have to be handled in an integrated system.

(2) External schema express the view that the user has of the data. A specific use for a specific application has a specific request for specific data. External schemas reflect those specificities. Two problems arise: how can those external schemas be mapped into the conceptual schema, in order to turn the specific view into the general schema; and how can the user express requests related to the external schemas.(Raper and Rhind 1990). To some extent the user would like to have no need of knowledge of how

the DBMS works and also to be able to express requests in his own language.

For those GDBs which are nationwide, the only request is for collecting a thematically and geographically restricted amount of information.(Salgé and Sclafer 1988, Bickmore and McCullagh 1989). The resulting data set is then transferred to a specific system where an interactive dialogue with the geographic data is performed.

(3) Internal schema express from the computer point of view where the data are to be stored and how the software is to have access to it. The main problem which arises from that level of software is the question of how to improve the efficiency of data retrieval. Further research will then be needed to choose and implement the most efficient way of storing data in a GDB (Goodchild 1989). Is there any parallelism algorithm which produces faster data retrieval (geometric keys)?

GDBs, due to their nationwide characteristics, raise other problems which have not yet been entirely solved. The huge amount of data (100 Gbits of highly structured data are currently quoted), give rise to research topics that computer scientists should examine. (Rhind 1988). As part of the historical patrimony of a country, national GDBs have to reflect the evolution of the countryside with time and supply a more abstract view of a portion of history. Next-generation DBMSs must then cope with the multiple representations of geographic objects in time and space. (Langran 1989, Guptill 1990, Salgé, 1990). In the future those kinds of GDB must be able to answer questions such as 'What data were available at time T1 about a given area at time T2' or 'What was the general evolution of a given area between time T3 and T4'? Before DBMSs can answer such questions they must be able to solve the problem of updating a GDB, that is to integrate updated knowledge describing a given area ; objects which are part of the semantic view of the Digital Geographic Information (DGI) have to be correctly identified and unequivocal identifiers have to be found.

As quality information and data about fuzzy geographic objects will be gradually captured, the DBMS of the future will be able to handle and retrieve them.

In future, objects of the real world, so-called features in SDTS Terminology, will be depicted through texts and images. As the next generation of GDBs will be to some extent object-oriented, the DBMS will handle in an integrated manner, texts, images and geographic information! (Boursier 1990). Here is the concept of an intelligent database which integrates various technologies (object-oriented, expert system, hypermedia...) (Bennis 1990).

DATA POST PROCESSING

Data from a nationwide GDB will be available to users who may not belong to the producer's organization. The user requirements will be diverse and imply to some extent a customization of the data set. Two kinds of customization will have to take place — internal and external customization.

Internal customization for a given data set consists of calculations or analyses inside the data set itself without any external data being involved. Among these are the manipulation of the internal topology. One can define a geometric topology where the data are considered to be immerged in a Euclidean plane, and a semantic topology where objects are related according to a specific network organization equals processing graphs, manipulating graphs and algorithms on graphs have to be re-invented with respect to DGI spatial characteristics.(Pressense and Salgé 1987, Raper and Rhind 1990). Data sets from a GDB may need to be vertically and horizontally integrated: separate layers will have to be combined by respecting their deep structure. On other hand, as thematic extraction from a GDB may define artefacts due to the mapping of the conceptual schema into the external schema, cleaning algorithms will be needed to erase them. Combination of layers from a GDB are also possible which will make it possible to derive composite information, eg contour lines plus hydrography ridges and valleys.(Guptill 1989, Falcidieno 1990). A knowledge-based analysis of the data gathered in a nationwide database may lead to derived databases which fulfil the user's requirements.

Quality information also has to be used and its visualization, and use in calculation are also part of data customization. In the same way time-information will be used for evolution analysis. (MARI 90 1990).

External customization means that external data can be merged with data from a GDB, for the same kinds of use. Thus registration problems will have to be solved and 'expert systems' will be useful for overcoming the lack of user awareness of such problems.

DATA EXPLOITATION

Data exploitation relates to reporting on the data in the GDB. This aspect includes cartographic and graphic outputs (on paper as well as on displays) and the generation of 'files' to be exchanged — the latter will require a discussion of the legal aspects of exchanging data which is not a research concern for geographers (perhaps lawyers will have to consider that aspect). (Chorley 1988).

According to map definition in a matrix, where direct and indirect visibility are the columns and hard copy and soft copy are the lines, cartographic and graphic outputs belong to real map and virtual map type 1 while the exchange of data relates to virtual map type 3 (and to some extent virtual map type 2). Those two types of problems raise different issues.(Moellering 1984).

As an output of a nationwide GDB, real maps will remain a specific problem. Generally speaking, GDBs are designed to fulfil requirements at specific scales. Here scales do not relate to any graphic ratio, but they express the level of abstraction by which the world is depicted. For example, a GDB designed for 1:25 000 allows the drawing of topographic maps from 1:10 000 to 1:50 000 using only a few specific processes. However no one can find realistic ways to derive cartographic outputs at 1:100 000 or 1:200 000.(Müller 1990). This operation is strongly related to generalization which is a well-known process in cartography. The classical way of performing such an operation relies on a set of rules which have not been totally established as they have been transmitted orally by generations of draughtsmen. Generalization is based on three main principles (Pernot 1989):

(1) Selection of the relevant objects to be featured and their relevant attributes.
(2) Definition of the object to be generalized as they relate to some extent to a specific conflict for a good understanding of the map to be drawn.
(3) Performing the generalization which is not only a geometric displacement but also a shape transformation with characteristic exaggeration and the caricature of some sets of objects.

The initial research will be to establish standards for the selection of objects for a given cartography, assumption being made that the relevant legend has been previously defined (an 'expert-system' for cartographic design has to be developed) (Pressense and Salgé 1987). The set of rules (more than 3000) have to be defined, and can be isolated by the analysis of specific questions put to expert cartographers and through the examination of maps and those from which they are derived.(Müller 1990). Important characteristics of this process are that all the components of the map have to be considered and treated simultaneously and that the method used for a given map will not be directly applicable to the next map. This is a research sector where expert systems or neuronal networks may apply. One can also relate to that issue the automatic name placement process and the setting-up of cartographic databases which keep track of the entire process of mapping from a GDB.

Virtual map type 2 (displays on a CRT) raises similar issues as the possibility of zoom (in and out) and panelling might be compared to some extent to the generalization process. Meanwhile, other themes are also to be considered which take advantage of the ability of computers to display quick, versatile and volatile images on a CRT. Unfortunately, technological improvements in graphic hardware far surpass developments in understanding the cognitive and perceptual mechanisms by which spatial patterns are identified and interpreted (Buttenfield and Ganter 1990). As Bertin's graphic semiology applies to cartographic drawing, (Bertin 1983) there is a need to apply that theory to CRT displays. Other types of representation will have to be derived from a GDB such as perspective views, the combination of line elements from a GDB with images (space maps), simulations (evolution of given phenomena). Links between GIS and image synthesis systems will also have to be realized and developed.(Salgé 1989).

Quality is often synonymous with certainty. However, data quality slants interpretation of displays (real map or virtual map type 2). Display of the quality of a derived map is an important issue which makes it possible to know how relevant is the displayed information. The quality of a decision depends on the quality of the map which is used and the display of quality information may minimize mistakes in the decision-making process.(NCGIA 1989, MARI 1990). There is another important issue: how to communicate quality information to the user so that he can use the data appropriately.

As an output of a nationwide GDB, virtual maps will have to be exchanged. One of their main aims is to disseminate digital geographic data. When a data set is extracted from a GDB, it will have to be processed into an exchange format. Assuming that standards on exchange format will have been defined, there remains the issue of transforming the data set from the format based on the conceptual data schema of the GDB to the format of the standard and its conceptual model. The first step is to interface the models before interfacing the data themselves. After a given data set has been delivered to a consumer, the GDB will continue its updating process and finally the question of delivering updated versions to the same user will arise. As he will probably have added specific data to the first data set, it is important to deliver updated versions which relate to the previous ones. Being able to trace various versions over a period of time will then become another challenge. Adding methods to objects or delivering change – only files are solutions which should be explored (Moellering 1990, Salgé 1990, Sowton 1990).

Exchanging messages in a standardized way so that a consumer can update his own database is another issue which implies taking advantage of recent advances in Electronic Data Interchange (EDI).

ORGWARE AND OPERATIONALITY

As previously stated, orgware and operationality account for some 40 % of the costs related to the development of a nationwide database. That leads to processes which have to be as user-friendly as possible. This raises the general topic of the human interface (NCGIA 1989). Taking into account the fact that the best user interface is the one which relies on the habits of the operators, one can try to study their knowledge and their behaviour. Nevertheless, a pragmatic approach needs to be followed; the GIS and other tools used for developing databases will have to offer some toolkits which enable a programmer to adapt the operator interface to the expertise of future operators (Burns and Henderson 1989, Goodchild 1990).

As the different applications behind the development of a GDB cannot be found from a single manufacturer, and as one must connect various systems which make it possible to exchange data without any degradation, one can define an 'internal' exchange format. This mechanism has to operate with the aim of being as quick as possible, and with the constraints of preserving all the semantic, topologic, and metric information in any data set. The problem of interfacing the various components of the GBD system is slightly different from the general problem of exchange standard because the various subsystems are well defined in the first case and thus the interface can be specified (Salgé et al. 1990). Organizational aspects also concern the configuration definition which makes it possible for the GBD to exist and the only possible approach then is to follow computer science research and try to take advantage of its results.

Organizational aspects also concern the human ability to face new aspects of digital geographic information. Developing a nationwide database is so costly an issue that it cannot be handled by a single person! Teams composed of some 100 people for several years are often quoted and thus all the processes will have to rely on execution staff. This means that teaching those operators is an issue which will have to be faced and an efficient learning process will have to be set up (NCGIA 1989).

As the development of a nationwide GDB aims at meeting, to some extent, the future needs of the users, another challenge will be to raise the awareness of what is feasible. Training users is an important issue which faces any producer and research will be needed to improve those educational aspects (Guptill 1990).

CONCLUSION

Producers of geographic information are generally governmental organizations. The role of producing maps was their main responsibility. They entered the world of digital geographic information via digital cartography. The last decade pushed them towards the concept of databases and the fact that their job is to deliver map data as well as digital geographic data. Gradually the semantic structure of the data has imposed itself as a key problem in fulfilling user requirements. For them this evolution looks more like a revolution!

Assuming that these trends towards user-defined products will be digested by those institutions, the next decade will be one of flexibility; in short, the research agenda for the development of a geographic database is aimed at adding this flexibility. Research leading to a better automation of all the processes surrounding the database, from data capture to their use, must be undertaken with the aim of having those systems operated by operators with only a limited scientific background.

GISs and consequently GDBs have a 'considerable scientific knowledge base' and that trend seems to be increasing as research continues. However, most of this knowledge has to be hidden from users, even if they are operators or engineers who employ GIS and GDB data for achieving their work.

PRACTITIONER'S SUMMARY

Present achievements allow cartographic organizations to build nationwide databases. With few exceptions, those databases are rather cartographic ones, even though most of the countries are undertaking studies and projects for real geographic databases. They are not scaleless as they are all built round a specific 'pivot' scale (ie for a narrow range of scales). One can hardly specify existing GDBs as seamless.

Data capture processes are set-up and rely mainly on teams of operators. The amount of automation is such that digital mapping programs are long term or contracted (and thus expensive). Solutions for increasing the current running rate are gradually set-up with specific attention paid to quality. No operationally suitable mechanism exists for updating nationwide databases.

Existing nationwide GDBs are built on specific softwares which try to take the best of existing DBMS. Meanwhile one can hardly say that data handling and data management systems are standardized and their efficiency is not that great.

Data post-processing and data exploitation mechanisms are currently rather producer oriented and information customization, for a specific user, is often limited to a mere extraction of all available data for a set of 'sheets'.

One of the main concerns is to raise user awareness of what is or is not possible using data from the GDB. Operator education and training are concerns which are actually solved in a pragmatic way.

Exchange mechanisms function erratically as few nationwide standards are in operation. National transfer formats are progressively being introduced and will lead to an organized and standardized way of exchanging geographic data (Moellering, Chapter 7).

PROJECTIVE SUMMARY

This projective summary takes the form of a clearly itemized research agenda arranged by decreasing priority according to the previous logic.

Data content definition

—data conceptual model
—database designer tools for specification writing
—feature-based spatial data model design
—international data conceptual schema
—scaleless–seamless databases and history
—systems and methods hiding the DCM/DCS design complexity

Data capture and updated data capture

—quality issues
—operational and robust data capture processes
—automatic geographic data extraction from digital images (scanned map, remote sensing images, digital air survey)
—semantic information capture based on GIS (with rules)
—capture oriented data structures
—use of GPS for geographic database
—efficient knowledge-based system for data capture.

Data handling and data management

—efficient DBMS for Geographic Databases
—mapping external schema (user query) into the conceptual schema
—efficient way of storing data
—handling huge amount of data and time related data
—intelligent update of GDB
—intelligent databases

Data post-processing

—cleaning algorithms
—derived databases and deductions from databases
—quality information processing
—horizontal and vertical integration
—efficient topologic operators
—expert systems for registration of heterogeneous data

Data exploitation

—real map drawing; symbolization, generalization, name placement
—real map design from GDB
—use of neuronal network for map generation
—graphic semiology on CRT display, cartographic image synthesis
—quality assessment communication to users
—exchange mechanisms issues; change only, methods attached to objects

Orgware and operationality

—human interface adapted to operator skill
—internal data exchange mechanisms equals interfacing GISs
—tools for teaching and organizing operators
—tools for training the users

REFERENCES AND BIBLIOGRAPHY

ATKIS German Specification AdV, 1990

Aangeenbrug R, Schiffman Y, (1987) The research agenda. In: *Proc. of Int. GIS Symp.*, Washington DC (USA)

Abiteboul S, Grumbach S, (1987) Bases de donneés et objets structurés. *Technique et Science Informatiques* **6** (5): 383–404

Abiteboul S, Beerl C, (1988) On the power of languages for the manipulation of complex objects. Rapport de recherche, INRIA

Adiba M, Ngoc B Q, (1986) Historical multimedia databases. In: *Proc. of the 12th Int. Conf. on Very Large Data Bases, Kyoto (Japan)*, 25–28 Aug. 1986

Algrain P, (1990) Vers une normalisation des formats d'échange d'informations géographiques numériques. In: *Actes de MARI* 90

Anderson S, (1987) The Swedish Land Data Bank *Int. J. of Geographical Information Systems* **1** (3)

Anderson T L, Ecklund E F Jr, Maier D, (1986) PROTEUS: Objectifying the DBMS user Interface. In: *Proc. of Int. Workshop on Object-Oriented Database Systems*, Pacific Grove (California, USA), 23–26 Sept. 1986: 133–47

Anselin L, (1989) What is special about spatial data? Alternative perspective on spatial data analysis. Technical Paper 89–4, NCGIA

Bancilhon F, Barbedette G, Benzaken V, Delobel C, Gamerman S, Lecluse C, Pfeffe P, Richard P, Velez F, (1988) The design and implementation of 02. An object-oriented database system. In: *Proc. of the OODBS II Workshop*, Bad Munster (RFA), Sept. 1988

Barwinski K, Bruggemann H, (1986) Development of digital cadastre and topographic maps: requirements, goals, and basic concepts. In: *Proc. of the Int. Conf. on the Acquisition, Management and Presentation of Spatial Data* (*Auto-Carto London*), London, Sept. 14–19, 1986, **2** 76–85

Batory D S, Barnett J, Garza J F, Smith K, Tsukuda K, Twichell C, Wise T, (1986) GENESIS: A reconfigurable database management system, Tech. Report, University of Texas

Becker L, Guting R H, (1989) Rule-based optimization and query processing in an extensible geometric database system, Forschungsberichte ders Fachbereichs Inf, Universität Dortmund

Bennis K, Freville M, (1990) Etude et implémentation d'un index spatial dans un SGBD cartographique In: *CARTAO*'90, Paris (France), Oct. 1990

Bennis K, David B, Quilio I, Viémont Y, (1990*a*) GéoTROPICS: Database support alternatives for geographic applications In: *Proc. of the 4th Int. Symp. on Spatial Data Handling*, Zürich (Switzerland), 23–27 July 1990

Bennis K, David B, Quilio l, Viémont Y, (1990*b*) Un SGBD cartographique basé sur un modèle relationnel étendu In: *CARTAO*'90, Paris (France), Oct. 1990

Bernstein P A, Dayal U, DeWitt D J, (1988) Al, Future directions in DBMS research In: *The Laguna Beach Workshop of the Int. Computer Science Institute*, 4–5 Feb. 1988

Bertin J, (1967) *Sémiologie Graphique*, Gauthiers Villars

Bertin J, (1983) *The Semiology of Graphics*, University of Wisconsin

Bianchin A, Jamet O, (1990) Measuring the accuracy of the land-use content of a GIS In: *Libro De Comunicaciones*, *Euro-Carto VIII*, Palma de Mallorca (España), 25–28 April 1990, Talleres del Servicio Geogràficio del Ej

Bickmore, D, McCullagh, M, (1989) Quality assurance for global databases. In: *Auto Cartog.*, Baltimore, USA

Borgida A, Mitchell T, Williamson K, (1986) Learning improved integrity constraints and schemas from exceptions in databases and knowledge bases In: *On Knowledge Base Management Systems*: 259–86

Bouillé F, (1989) Integrating expert system and database for G.I.S. In: *Proc. of SORSA'89 Colloquium*, University of Maryland, USA, March 29–31, 1989

Boursier P, (1990) Euromap: a CD ROM based digital cartographic atlas of Europe In: *Proc. of the First European Conference on Geographical Information Systems* (*EGIS*'90), Amsterdam (The Netherlands), April 10–13, 1990

Boursier P, Jeansoulin R, (1986) Intégration de données cartographiques et d'images dans Ies systèmes d'information géographique In: *Actes du deuxième colloque Image*: *traitement*, *synthèse*, *technologie et applications*, Nice (France), 21–25 April 1986, **1**: 47–51

Boursier P, Jeansoulin R, (1986) Bases de données images et systèmes d'information géographique In: *Actes des journées francophones sur l'informatique, bases de données et bases de connaissances*, Grenoble (France), Jan. 1986

Bouzeghoub M, (1986) SECSI: Un système expert de conception en systèmes d'information, Thèse de doctorat, Université Paris VI, 1986

Bracken I, Webster C, (1989) Towards a Typology of GIS, *Int. J. of Geographical Information Systems*, **3** (2): 137–52

Brassel K, (1990) Overview of Spatial Information Systems In: *Photogrammetry and Land Information Systems*, Lausanne (Switzerland), March 6–17, 1989: 13–21

Broome F R, (1984) TIGER Preliminary Design and Structure Overview: The Core of the Geographic Support System for 1990 In: *Proc. of Annual Meeting of the Association of the American Geographers*, Washington D.C. (USA), 1984

Bruegger B P, Frank A U, (1989) Hierarchies Over Topological Data Structures In: *Tech. papers of ASPRS/ACSM Annual Convention*, Baltimore (Maryland, USA), April 2–7, 1989: 137–45

Brüggemann H, (1990) ATKIS and European Road Database Interrelations between Producers and Users of Topographic Data In: *Libro De Comunicaciones, Euro-Carto VIII*, Palma de Mallorca (España), 25–28 April 1990, Talleres del Servicio Geogràficio del Ej

Buisson L, (1989) Reasoning on Space with Obiect-Centred Knowledge Representations In: *Proc. of the Symp. on the Design and Implementation of Large Spatial Databases*, Santa Barbara (California, USA), July 1989, Lecture Notes in Computer Science no 409

Burns T, Henderson I, (1989) Educational and Training in GIS: the view from ESRI In: *Proc. of the Ninth Int. Symp. on Computer-Assisted Cartography* (*Auto-Carto* 9), Baltimore (Maryland, USA), April 2–7, 1989

Buttenfield B P, Ganter J H, (1990) Visualization and GIS: What Should We See? In: *Proc. of the 4th Int. Symp. on Spatial Data Handling*, Zürich (Switzerland), July 23–27, 1990, Vol. 1: 307–16

Carey M J, DeWitt D J, Frank D, Graefe G, Muralikrishna M, Richardson J E, Shekita E J, (1986) The Architecture of the EXODUS Extensible DBMS In: *Proc. of Int. Workshop on Object-Oriented Database Systems*, Pacific Grove (California, USA), September 23–26, 1986: 52–65

Carey M J, DeWitt D J, Vandenberg S L, (1988) A Data Model and Query Language for EXODUS In: *Proc. of ACM-SIGMOD Int. Conf. on Management of Data*, Chicago (Illinois, USA), June 1–3, 1988, SIGMOD Record: 413–23

Charlwood G, Moon G, Tulip J, (1987) Developing a DBMS for geographic information: a review In: *Proc. of the Eighth Int. Symp. on Computer-Assisted Cartography* (*Auto-Carto* 8), Baltimore (Maryland, USA), March 29–April 3, 1987

Lord Chorley, (1988) Some Reflections on the Handling of geographical Information, *Int. J. of Geographical Information Systems* **2** (1): 3–10

Chrisman N R, (1987) Design of Geographic Information Systems based on Social and Cultural Goals, *Photogrammetric Engineering & Remote Sensing*, **53**, (10) 1987, : 1367–70

Chrisman N R, (1989) Error in categorical map: Testing versus Simulation In: *Proc. of the Ninth Int. Symp. on Computer-Assisted Cartography* (*Auto-Carto* 9), Baltimore (Maryland, USA), April 2–7, 1989

Damier C, (1989) OMEGA: Un SGBD multimédia orienté objet pour Ies applications géographiques, Thèse de doctorat, Université Grenoble 1

Damier C, Defude B, (1988) Un modèle de données pour les informations géographiques In: *Actes des quatrièmes Journées Bases de Données Avancées* (BD3), Bénodet (France), 17–20 May 1988

David B, (1988) Le Modèle SPATlAREL: une extension du modèle relationnel pour gérer les données spatiales surfaciques In: *Actes des quatrièmes Journées Bases de Données Avancées* (BD3), Bénodet (France), 17–20 May 1988: 73–93

David B, (1988) Map is Better than Graph to Store Topology In: *Proc. of Euro-Carto Seven*, Enschede (The Netherlands), September 20–22, 1988, ITC Publication

David B, (1989) External Specifications for the Cartographic DBMS, Technical Report 22, TROPICS project

David B, (1990) Data Structures Alternatives for Very Large Spatial Databases In: *Proc. of SORsA'90 Colloquium*, Freiburg (Germany), July 28–29, 1990

Denègre J, (1987) Troupel F, Le plan de base Topo-Foncier, *Le Géomètre*, (6): 20–21

Denègre J, (1988) L'information géographique, *Le Géomètre*, **8**

Destival I, (1987) Recherche automatique des réseaux linéaires sur les images SPOT, Bulletin de Ia S.F.P.T, no. 105: 5–16

Drummond, J, (1988) Fuzzy subset theory applied to environmental planning, GS, EUROCARTO 7, TTC publication no. 8

Dueker K J, (1987) Multipurpose Land Information Systems: Technical. Economic and Institutional Issues, *Photogrammetric Engineering & Remote Sensing*. **53** (10): 1361–6

Dufourd J F, (1989) Gross C, Spehner J-C, A Digitisation Algorithm for the Entry of Planar Maps In: *Proc. of the 7th Int. Conf. on Computer Graphics*, Leeds (UK), June 27–30, 1989

Van Eck J W, Uffer M, (1990) A Presentation of System 9 in: *Photogrammetry and Land Information Systems*, Lausanne (Switzerland), March 6–17, 1989: 139–78

Egenhofer M J, (1990) Extending SQL for Cartographic Output

Egenhofer M J, Frank A U, (1988) Towards a Spatial Query Language: User Interface Considerations In: *Proc. of the 14th Int. Conf. on Very Large Data Bases*, Los Angeles (California, USA), 1988: 124–33

Egenhofer M J, Frank A U, (1988) Designing Object-Oriented Query Languages for GIS: Human Interface Aspects In: *Proc. of the Third Int. Symp. on Spatial Data Handling*, Sydney (Australia), August 17–19, 1988: 79–97

Egenhofer M J, Frank A U, (1990) LOBSTER: Combining Al and Database Techniques for GIS, *Photogrammetric Engineering & Remote Sensing*, **56** (6): 919–26

Egenhofer M J, Herring J R, (1990) A Mathematical Framework for the Definition of Topological Relationships In: *Proc. of the 4th Int. Symp. on Spatial Data Handling*, Zurich (Switzerland), July 23–27, 1990, Vol. 2: 803–13

Egenhofer M J, Frank A U, Jackson J P, (1989) A Topological Data Model for Spatial Databases In: *Proc. of the Symp. on the Design and Implementation of Large Spatial Databases*, Santa Barbara (California, USA), July 1989, Lecture Notes in Computer Science no 409

Falcidieno B, (1990) A feature based approach to terrain surface approximation In: *Proc. of the 4th Int. Symp. on Spatial Data Handling*, Zürich (Switzerland), July 23–27, 1990

Faloutsos C, Sellis T, Roussopoulos N, (1987) Analysis of Object Oriented Spatial Access Method In: *Proc. of ACM-SIGMOD Annual Conf.*, San Francisco (California, USA), May 27–29, 1987, SlGMOD Record: 426–39

Faust N L, (1987) Automated Data Capture for Geographic Information Systems: A Commentary, *Photogrammetric Engineering & Remote Sensing*, **53**, (10): 1389–92

Fegeas R F, Pearsall R A, (1984) Topological Structuring of Raster-Scanned Line Map Data In: *Tech. papers of ASP/ACM Annual Convention*, Washington D.C. (USA), March 11–16, 1984

Forest A R, (1987) Computational Geometry and Software Engineering: Towards a Geometric Computing Environment In: *Techniques for Computer Graphics*: 23–37

Frank A U, (1982) Map Query: Database Query Language for Retrieval of Geometric Data and their Graphical Representation, *Computer Graphics*, **16** (3): 199–207

Frank A U, (1984) Requirements for Databases Systems Suitable to Manage Large Spatial Databases In: *Proc. of the Int. Symp. on Spatial Data Handling*, Zürich (Switzerland), August 20–24, 1984: 38–60

Frank A U, Kuhn W, (1986) Cell Graphs: A Provable Method for the Storage of Geometry In: *Proc. of the Second Int. Symp. on Spatial Data Handling*, Seattle (Washington, USA), July 5–10, 1986: 411–36

Frank A U, (1988) Requirement for a Database Management System for a GIS, *Photogrammetric Engineering & Remote Sensing*. **54** (11)

Franklin W R, Wu P Y F, (1987) A Polygon Overlay System in Prolog In: *Proc. of the Eighth Int. Symp. on Computer-Assisted Cartography* (*Auto-Carto* 8), Baltimore (Maryland, USA), March 29–April 3, 1987: 97–106

Franklin W R, (1984) Cartographic Errors Symptomatic of Underlying Algebra Problems In: *Proc. of the Int. Symp. on Spatial Data Handling*, Zürich (Switzerland), August 20–24, 1984: 190–208

Freeston M W, (1989) A Well-Behaved File Structure for the Storage of Spatial Objects In: *Proc. of the Symp. on the Design and Implementation of Large Spatial Databases*, Santa Barbara (California, USA), July 1989, Lecture Notes in Computer Science no 409

Fréville M, (1990) Etude et implementation d'une structure d'index spatial dans un SGBD cartographique, Mémoire d'ingénieur, C.N.A.M. Paris

Gahegan M N, Roberts S A, (1988) An Intelligent object-Oriented Geographical Information System, *Int. J. of Geographical Information Systems*, **2** (2): 101–10

Garcia L, Rodriguez J A, Razola J F, (1990) Towards a GIS in the Instituto Geographico Nacional of Spain In: *Libro De Comunicaciones, Euro-Carto VIII*, Palma de Mallorca (España), 25–28 April 1990, Talleres del Servicio Geogràficio del Ej

Gardarin G, (1982) Bases de données: les systèmes et leurs langages, Eyrolles

Gardarin G, Simon E, (1987) Les systèmes de gestion de bases de données déductives, *Technique et Science Informatiques*, **6** (5)

Gardarin G, Valduriez P, (1985) Bases de données relationnelles: analyse et comparaison de systèmes, Eyrolles

Gardarin G, Valduriez P, (1989) *Relational Databases and Knowledge Bases*, Addison Wesley Pub. Co

Gardarin G, Valduriez P, (1990) SGDB avancés: Bases de données objets. déductives. réparties, Eyrolles

Gardarin G, Pucheral P, Thévenin J-M, (1988) SteHen H, Overall Architecture of the GEODE Extensible DBMS Kernel In: *Proc. of the 5th ESPRIT* Conf., Bruxelles (Belgium), October 1988

Gardarin G, Cheiney J-P, Kieman G, Pastre D, Stora H, (1989) Managing Complex Objects in an Extensible Relational DBMS In: *Proc. of the 15th Int. Conf. on Very Large Data Bases*, Amsterdam (The Netherlands), August 22–25, 1989

Gardels K D, (1987) The Expert Geographic Knowledge System Applying Logical Rules to Geographical Information In: *Proc. of the Eighth Int. Symp. on Computer-Assisted Cartography* (*Auto-Carto* 8), Baltimore (Maryland, USA), March 29–April 3, 1987

Goodchild M F, (1987) Towards an Enumeration and Classification of GIS functions In: *Proc. of Int. GIS Symp.*, Arlington (Va, USA), November 1987, Vol. 2: 67–77

Goodchild M F, (1989) Optimal Tiling for Large Cartographic Databases In: *Proc. of the Ninth Int. Symp. on Computer-Assisted Cartography* (*Auto-Carto* 9), Baltimore (Maryland, USA), April 2–7, 1989: 444–51

Goodchild M F, (1990) Spatial Information Science (Keynote Address) In: *Proc. of the 4th Int. Symp. on Spatial Data Handling*, Zürich (Switzerland), July 23–27, 1990, Vol. 1: 3–14

Greene D, (1989) An Implementation and Performance Analysis of Spatial Data Access Methods In: *Proc. of Fifth Int. Conf. on Data Engineering*, Los Angeles (California, USA), February 6–10, 1989: 606–15

Günther O, Wong E, (1987) A Dual Source Representation for Geometric Data In: *Proc. of the 13th Int. Conf. on Very Large Data Bases*, Brighton (England), September 1– 4, 1987: 501–6

Günther O, (1989) The Design of the Cell Tree: An object-Oriented Index Structure for Geometric Databases In: *Proc. of Fifth Int. Conf. on Data Engineering*, Los Angeles (California, USA), February 6–10, 1989

Guptill S C, (1986) A New Design for the U.S. Geological Survey's National Digital Cartographic Data Base In: *Proc. of the Int. Conf. on the Acquisition, Management and Presentation of Spatial Data* (*Auto-Carto London*), London (UK), September 14–19, 1986, Vol. 2: 10–18

Guptill S C, (1987) Desirable Characteristics of a Spatial Data Base Management System In: *Proc. of the Eighth Int. Symp. on ComputerAssisted Cartography* (*Auto-Carto* 8), Baltimore (Maryland, USA), March 29–April 3, 1987: 278–81

Guptill S C, (1989) Speculations on Seamless, Scaleless, Cartographic Databases In: *Proc. of the Ninth Int. Symp. on Computer-Assisted Cartography* (*Auto-Carto* 9), Baltimore (Maryland, USA), April 2–7, 1989

Guptill S C, (1989) Inclusion of Accuracy Data in a Feature Based, Object Oriented Data Model In: *Accuracy of Spatial Databases*: *Specialist Meeting Report*, Montecito (California, USA), Dec. 13–16, 1988, Technical Paper 89–1

Guptill S C, (1990) AL, An Enhanced DLG design, USGS circular 1048, USGS

Guptill S C, (1990) Multiple Representations of Geographic Entities through Space and Time In: *Proc. of the 4th Int. Symp. on Spatial Data Handling*, Zurich (Switzerland), July 23–27, 1990, Vol. 2: 859–68

Guptill S C, Claire R W, (1982) Spatial Operators for Selected Data Structures In: *Proc. of Auto-Carto V*, 1982

Guptill S C, Starr L E, (1988) The Future of Cartography in the Information Age In: *Computer-Assisted Cartography Research and Development Report*, 1984: 1–15
Guptill S C, Fegeas R F, Domaratz M A, (1988) Designing an Enhanced Digital Line Graph In: *Tech. Papers of ACSM–ASPRS Annual Convention*, St. Louis (Missouri, USA), March 13–18, 1988, Vol. 2: 252–61
Güting R H, (1989) Gral: An Extensible Relational Database System for Geometric Applications In: *Proc. of the 15th Int. Conf. on Very Large Data Bases*, Amsterdam (The Netherlands), August 22–25, 1989
Haywood P, (1988) From Digital maps to Structured Data and Database Management at Ordnance Survey In: *Proc. of Euro-Carto Seven*, Enschede (The Netherlands), September 20–22, 1988, ITC Publication: 115–25
Haywood P, (1990) An Integrated Production and Business Strategy In: *Libro De Comunicaciones, Euro-Carto VIII*, Palma de Mallorca (España), 25–28 April 1990, Talleres del Servicio Geogràficio del Ej
Healey R G, Desa G B, (1989) Transputer-Based Parallel Processing for GIS Analysis: Problems and Potentialities In: *Proc. of the Ninth Int. Symp. on Computer-Assisted Cartography* (Auto-Carto 9), Baltimore (Maryland, USA), April 2–7, 1989: 90–9
Herring J R, (1987) TIGRIS: Topologically Integrated Information System In: *Proc. of the Eighth Int. Symp. on Computer-Assisted Cartography* (*Auto-Carto* 8), Baltimore (Maryland, USA), March 29–April 3, 1987: 282–91
Herring J R, (1990) The Definition and Development of a Topological Spatial Data System In: *Photogrammetry and Land Information Systems*, Lausanne (Switzerland), March 6–17, 1989: 57–70
Herring J R, Larsen R C, Shivakumar J, (1988) Extensions of the SOL Query Language to Support Spatial Analysis in a Topological Data Base In: *GIS/LIS*'88 *Proceedings*, San Antonio (Texas, USA), Nov. 30–Dec. 2, 1988: 741–50
Hoffmann C, (1989) The Problems of Accuracy and Robustness in Geometric Computation, *lEEE Computer*, **22** (3)
Ilg M, (1990) Knowledge-based Interpretation of Road Maps In: *Proc. of the 4th Int. Symp. on Spatial Data Handling*, Zurich (Switzerland), July 23–27, 1990, Vol. 1: 25–34
Imai H, (1986) Asano T, Efficient Algorithms for Geometric Graph Search Problems, *IAM J. of Computing*, **15** (2): 478–94
Jackson M J, Mason D C, (1986) The development of integrated geo-information systems, *Int. J. of Remote Sensing*, **7** (6): 723–40
Jackson J P, (1989) Algorithms for Triangular Irregular Networks based on Simplicial Complex Theory In: *Tech. papers of ASPRS/ACSM Annual Convention*, Baltimore (Maryland, USA), April 2–7, 1989
Jasinski M J, (1990) The Comparison of Complexity Measures for Cartographic Lines, Technical Paper 90–1, NCGIA
Jong De W M, Lucas A H G, (1989) Development of a large National Geographic Database in the Netherlands In: *Proc. of SORSA*'89 *Colloquium*, University of Maryland (Maryland, USA), March 29–31, 1989
Jungert E, (1984) Inference Rules in a Geographical Information System In: *Proc. of IEEE Workshop on Languages for Automation*, New Orleans (USA), Nov. 1984: 32–7
Kemp Z, (1990) An object-Oriented Model for Spatial Data In: *Proc. of the 4th*

Int. Symp. on Spatial Data Handling, Zürich (Switzerland), July 23–27, 1990, VoI. 2: 659–68

Kemper A, Wallrath M, (1987) An Analysis of Geometric Modelling in Database Systems, *ACM Computing Surveys*, **19** (1)

Kinnear C, (1987) The TIGER Structure In: *Proc. of the Eighth Int. Symp. on Computer-Assisted Cartography* (*Auto- Carto* 8), Baltimore (Maryland, USA), March 29–April 3, 1987: 249–57

Kirkpatrick D, (1983) Optimal Search in Planar Subdivisions, *SlAM J. of Computing*, **12** (1): 28–35

Kosko B, (1986) Fuzzy cognitive maps, *Int. J. Man–Machines Studies* (24): 65–75

Langran G, (1989) Accessing Spatiotemporal Data in a Temporal GIS In: *Proc. of the Ninth Int. Symp. on Computer-Assisted Cartography* (*Auto-Carto* 9), Baltimore (Maryland, USA), April 2–7, 1989: 191–8

Langran G, (1989) A Review of Temporal Database Research and its Use in GIS Applications, *Int. J. of Geographical Information Systems* **3** (3): 215–232

Langran G, Chrisman N R, (1988) A Framework for Temporal Geographic Information, *Cartographica*, **25** (3): 1–14

Laurini R, (1985) Graphics Information Systems Based on Peano Space Piling Curves In: *Proc. of Eurographics*, Nice (France), 1985

Lauzon J-P, McLaren R, Harwood C, (1985) Structure d'une base de données pour les systèmes d'information géographique: la démarche suivie pour le système 9, Le Géomètre, no. 11: 59–63

Lepaumier L-P, (1988) Bibliographie et comparaison de structures de données géométriques, Rapport interne, IGN-F

Lienhardt P, (1989) Subdivisions of Surfaces and Generalized Maps In: *Proc. of Eurographics*, 1989

Linnemann V, (1988) AL, Design and Implementation of an Extensible Database Management System Supporting User Defined Data Types and Functions In: *Proc. of the* 14*th Int. Conf. on Very Large Data Bases*, Los Angeles (California, USA), 1988: 294–305

Mainguenaud M, (1989) Un langage graphique pour l'interrogation de réseaux: application aux bases de données géographiques, Thèse de doctorat, Université Paris VI

Mainguenaud M, (1990) Is an Extended Relational Database System Powerful Enough to DeaI with Network Applications In:*Proc. of the First European Conference on Geographical Information Systems* (*EGIS*'90), Amsterdam (The Netherlands), April 10–13, 1990: 701–10

Manola F A, (1986) Orenstein J A, Toward a General Spatial Data Model for an Object-Oriented DBMS In: *Proc. of the* 12*th Int. Conf. on Very Large Data Bases*, Kyoto (Japan), August 25–28, 1986: 328–35

Marble D F, (1985) Geographic Information Systems: an overview In: *Proc. of Pecora* 9: *IEEE Int. Conf. on Spatial Information Technologies for Remote Sensing Today and Tomorrow*, Sioux-Falls (USA), Oct. 2–4, 1984

MARI 90, (1990) L'Information Géographique Numérique, Paris, 4–6 April 1990, CFC bulletin no.126, Dec. 1990

Mark D M, (1989) Working Bibliography on 'Languages of Spatial relations'. First Edition, Technical Paper 89–10, NCGIA

Masser l, (1988) The Regional Research Laboratory Initiative. A Progress Report, *Int. J. of Geographical Information Systems*, **2** (1): 11–22

McCallum D G, Queree C, Robins R G, (1990) PANDORA — Prototyping a Navigation Database of Road-Network Attributes In: *Libro De Comunicaciones, Euro-Carto VIII*, Palma de Mallorca (España), 25–28 April 1990, Talleres del Servicio Geogràficio del Ej

Mcllroy, A Note on Discrete Representation of Lines, *AT&T Technical Journal*, **64** (2): 481–90

Meixler D, Saalfeld A, (1985) Storing. Retrieving and Maintaining Information on Geographic Structures: A Geographic Tabulation Unit (GTUB) approach In: *Auto-Carto 7 Proc., Digital Representation of Spatial Knowledge*, Washington D.C. (USA), March 11–14, 1985: 369–76

Moellering H, (1984) Real Maps, Virtual Maps and Interactive Cartography In: *Spatial Statistics and Models*

Moellering H, (1986) A Review and Definition of 0– 1– and 2–Dimensional objects for Digital Cartography In: *Proc. of the Second Int. Symp. on Spatial Data Handling*, Seattle (Washington, USA), July 5–10, 1986: 616–27

Moellering H, (1986) Developing Digital Cartographic Data Standards for the United States In:*Proc. of the Int. Conf. on the Acquistion, Management and Presentation of Spatial Data* (*Auto-Carto London*), London(UK), September 14–19, 1986: 312–22

Moellering H, (1990) Introduction In: *ACl monograph on exchange standards*

Mohan L, Kashyap R L, (1988) An object-Oriented Knowledge Representation for Spatial Information, *lEEE Trans. on Software Engineering*, **14**(5): 675–81

Morehouse S, (1985) ARC/INFO: A Geo-Relational Model for Spatial Information In: *Auto-Carto 7 Proc., Digital Representation of Spatial Knowledge*, Washington D.C. (USA), March 11–14, 1985: 388–97

Morehouse S, (1989) The Architecture of ARC/INFO In: *Proc. of the Ninth Int. Symp. on Computer-Assisted Cartography* (*Auto-Carto* 9), Baltimore (Maryland, USA), April 2–7, 1989: 266–77

Morehouse S, (1990) The Role of Semantics in Geographic Data Modelling In: *Proc. of the 4th Int. Symp. on Spatial Data Handling*, Zürich (Switzerland), July 23– 27, 1990, VoI. 2: 689–98

Morrison J L, Callahan G M, Olsen R W, (1987) Digital Systems Development at the U.S. Geological Survey In: *Proc. of the Int. Cartographic Association Conference*, Morelia (Michoacan, Mexico), Oct. 12–21, 1987: 201–14

Motet S, (1986) Schéma relationnel pour une Base de Donnée Topographique In: *Proc. of Euro-Carto V*, 1986

Motet S, (1989) Traitement de l'information géographique, DIEA Sciences de l'info. géographique

Motet S, (1989) Etude préalable pour une loite à outils geographiques. PRC Base de Données INT Evry. 1989

Motet S, (1990) An Interface Using the Relational 'Plane' Domain In: *Proc. of the 4th Int. Symp. on Spatial Data Handling*, Zürich (Switzerland), July 23–27, 1990, Vol. 1: 450–4

Motet S, Couesnon T, Laurent D, (1984) 'Geomatic' a 3–D Graphic Relief Simulation System In: *Frontiers in Computer Graphics*

Muller J-C, (1990) Rule Based Generalization: Potentials and Impediments In:

Proc. of the 4th Int. Symp. on Spatial Data Handling, Zürich (Switzerland), July 23–27, 1990, Vol. 1: 317–37

NCGIA, (1989) The research plan of the National Center for Geographic Information Analysis, *Int. J. of Geographical Information Systems*, **3** (2): 117–36

Nierstrasz O, (1989) A Survey of object Oriented Concepts In: *Object-Oriented Concepts, Databases, and Applications*, ACM Press, Frontier Series: 3–21

Noronha V T, (1988) A Survey of Hierarchical Partitioning Methods for Vector Images In: *Proc. of the Third Int. Symp. on Spatial Data Handling*, Sydney (Australia), August 17–19, 1988: 185–200

Ooi B C, Sacks-Davis R, McDonell K J, (1989) Extending a DBMS for Geographic Applications In: *Proc. of Fifth Int. Conf. on Data Engineering*, Los Angeles (California, USA), February 6–10, 1989: 590–7

Van Oosterorn P, Van Den Bos J, (1989) An Object-Oriented Approach to the Design of Geographic Information Systems In: *Proc. of the Symp. on the Design and Implementation of Large Spatial Databases*, Santa Barbara (California, USA), July 1989, Lecture Notes in Computer Science no 409

Orenstein J A, (1986) Spatial Query Processing in an object- Oriented Database System In: *Proc. of ACM-SIGMOD, 1986 Int Conf on Management of Data*: 326–36

Orenstein J A, (1989) Strategies for Optimizing the Use of Redundancy in Spatial Databases In: *Proc. of the Symp. on the Design and Implementation of Large Spatial Databases*, Santa Barbara (California, USA), July 1989, Lecture Notes in Computer Science no 409

Orenstein J A, (1989) Redundancy in Spatial Databases in: *Proc. of the ACM SIGMOD Int. Conf. on the Management of Data*, Portland (Oregon, USA), May 31–June 2, 1989, SIGMOD Record, Vol. 18: 295–305

Pasquier B, (1987) Cartographie numérique des dornaines: structuration. modélisation. algorithmique, Thèse d'état, Université Paris VI

Pernot, M, (1989) Étude d'un système de réaction cartographique, Rapport de DEA, École Nationale du Ponts et Chausses, Paris

Peuquet D J, (1986) The Use of Spatial Relationships to Aid Spatial Database Retrieval In: *Proc. of the Second Int. Symp. on Spatial Data Handling*, Seattle (Washington, USA), July 5–10, 1986: 459–71

Pressensé L, Salgé F, (1987) Les bases de données localisées à l'IGN, *La jaune et la rouge*, **426** no. juin–juillet

Pullar D V, Egenhofer M J, (1988) Toward Formal Definitions of Topological Relations Among Spatial Objects In: *Proc. of the Third Int. Symp. on Spatial Data Handling*, Sydney (Australia), August 17–19, 1988: 225–42

Quilio I, Dupuis P, David B, Ambrosini J-M, (1989) Design of Architecture and Interfaces of GéoTROPICS, Technical Report 36, TROPICS Project

Raper J, Rhind D W, (1990) UGIX(A]: The Design of a Spatial Language Interface for a Topological Vector GIS In: *Proc. of the 4th Int. Symp. on Spatial Data Handling*, Zürich (Switzerland), July 23–27, 1990

Rhind D W, (1988) A GIS research agenda, *Int. J. of Geographical Information Systems*

Rhind D W, Mounsey H, (1988) The Chorley Committee and 'Handling Geographic Information' In: *Proc. of the Third Int. Symp. on Spatial Data Handling*, Sydney (Australia), August 17–19, 1988: 407–9

Robinson V B, (1987) Expert Systems applied to Problems in Geographical Information Systems In: *Proc. of the Eighth Int. Symp. on Computer-Assisted Cartography* (*Auto-Carto* 8), Baltimore (Maryland, USA), March 29–April 3, 1987

Rowey L A, (1986) A Shared object Hierarchy In: *Proc. of Int. Workshop on Object-Oriented Database Systems*, Pacific Grove (California, USA), September 23–26, 1986: 160–70

Saalfeld A, (1985) Lattice Structures in Geography in: *Auto-Carto* 7 *Proc., Digital Representation of Spatial Knowledge*, Washington D.C. (USA), March 11–14, 1985: 482–9

Sacks-Davis R, McDonell K J, Ooi B C, (1987) GEOQL—A Query Language for Geographic Information Systems In: *Proc. of Australian and New Zealand Association for the Advancement of Science Congress*, Townsville (Australia), August 1987

Salgé F, (1989*a*) Conditions for Dialogue Between Heterogeneous Geographic Information Systems In: *OCDE Coordinated Information Systems for Urban Functioning and Management*, Copenhagen (Denmark), Oct. 11–13, 1989

Salgé F, (1989*b*) Geographic, Cartographic, Located Data Exchange Formats. An overview of the existing solutions over the world In: *Proc. of the* 13*th Urban Data Management Symposium* (UDMS 89), Lisboa, May 31–June 2, 1989

Salgé F, (1989*c*) Le système d'information géographiques nationales de I'lGN in: *Actes de VISUDA*, Paris (France), 1989

Salgé F, (1990*a*) An ETF: fact or fiction In: *proc. of AM/FM Nordic Conf.*, Ronneby brun (Sweden), 1990

Salgé F, (1990*b*) La normalisation des échanges d'informations géographiques en France In: *ACI monograph on exchange standards*, 1990

Salgé F, (1990*c*)The National Geographic Information System of the French National Geographic Institute In: *Proc. of the First European Conference on Geographical Information Systems* (*EGIS*'90), Amsterdam (The Netherlands), April 10–13, 1990

Salgé F, (1990*d*) L'échange et Ia diffusion des données géographiques numériques: synthèse et perspectives Européennes In: *Troisièmes Journées de la Recherche du CNlG*, Paris (France), 22 May 1990

Salgé F, Piquet-Pellorce D, (1986) The I.G.N. Small Scale Geographical Database (1:100.000 to 1:500.000) In: *Proc. of the Int. Conf. on the Acquisition, Management and Presentation of Spatial Data* (*Auto-Carto London*), London (UK), September 14–19, 1986, Vol. 1: 433–46

Salgé F, Sclafer M-N, (1988) The IGN Cartographic Database From the Users' Needs to the Relational Structure In: *Proc. of Euro-Carto Seven*, Enschede (The Netherlands), September 20–22, 1988, ITC Publication: 51–65

Salgé F, Sclafer M-N, (1989) A Geographic Data Model Based on HBDS Concepts: The IGN Cartographic Data Base Model In: *Proc. of the Ninth Int. Symp. on Computer-Assisted Cartography* (*Auto-Carto* 9), Baltimore (Maryland, USA), April 2–7, 1989: 110–17

Salgé F, Sclafer M-N, Etienne P, Péné J, Priou J-Y, Richard D, (1990*a*) Interfacing GIS's In: *Libro De Comunicaciones, Euro-Carto VIII*, Palma de Mallorca (España), 25–28 April 1990, Talleres del Servicio Geogràficio del Ej

Salgé F, Sclafer M-N, Faad, (1990*b*) Real World Description Models: the IGN-F

BDTopo and BDCarto examples In: *AM/FM European Division*, Montreux (Switzerland), 1990

Salgé F, Sclafer M-N, Priou J Y, Etienne P, (1990*c*) Interfacing GIS In: *AM/FM European Division*, Montreux (Switzerland), 1990

Sarnet H A, Rosenfeld A, Shaffer C A, Webber R E, (1984) Use of Hierarchical Data Structures in Geographical Information Systems In: *Proc. of the Int. Symp. on Spatial Data Handling*, Zurich (Switzerland), August 20–24, 1984: 392–411

Sanderson R H, (1988) A practical Feasibility Study of the Implementation of a Large Geographical Database

Schek H J, (1990) Relational Database Concepts and Research Aspects to Cover Spatial Data Needs In: *Photogrammetry and Land Information Systems*, Lausanne (Switzerland), March 6–17, 1989: 23–34

Scholl M, Voisard A, (1989) Modeling of Thematic Maps: an Application to Geographic Databases In: *Actes des cinquièmes journées bases de données avancées*, Genève (Suisse), Septembre 1989

Sclafer M-N, Salgé F, (1987) Base de Données Cartographiques: spécifications du contenu et de la structure, IGN

Seeger B, Kriegel H-P, (1988) Techniques for Design and Implementation of Efficient Spatial Access Methods In: *Proc. of the 14th Int. Conf. on Very Large Data Bases*, Los Angeles (California, USA), 1988: 360–71

Shepherd J W, Masser I, Blakemore M, Rhind D W, (1989) The ESRC's Regional Research Laboratories: An Alternative Approach to the NCGIA In: *Proc. of the Ninth Int. Symp. on Computer-Assisted Cartography* (*Auto-Carto* 9), Baltimore (Maryland, USA), April 2–7, 1989: 764–74

Six H V, Widmayer P, (1988) Spatial Searching in Geometric Databases In: *Proc. of Fourth Int. Conf. on Data Engineering*, Los Angeles (California, USA), February 1–5, 1988: 496–503

Smith I, (1990) DGIWG exchange standards In: *ACI monograph on exchange standards*

Smith N, (1989) The Ordnance Survey Pilot Topographic Database since the Durham Symposium In: *Proc. of SORSA'89 Colloquium*, University of Maryland (Maryland, USA), March 29–31, 1989

Smith T R, Pazner M I, (1984) A Knowledge Based System for Answering Queries Concerning Geographic Objects In: *Proc. of Pecora 9: IEEE Int. Conf. on Spatial Information Technologies for Remote Sensing Today and Tomorrow*, Sioux-Falls (USA), Oct. 2–4, 1984

Smith T R, Peuquet D J, Menon S, Agarwal P, (1987) KBGIS-ll: A Knowledge-Based Geographical Information System, *Int. J. of Geographical Information Systems* **1** (2): 149–72

Smith T R, Menon S, Star J L, Estes J E, (1987) Requirements and Principles for the Implementation and Construction of Large-Scale Geographic Information Systems, *Int. J. of Geographical Information Systems***1**(1): 13–31

Sopena E, (1986) Réécriture d'hypercartes combinatoires et conception d'un langage pour Ia programmation d'algorithmes de graphes, Thèse de doctorat, U. de Bordeaux L

Sowton M, (1990) CERCO consideration concerning an ETF In: *ACl monograph on exchange standards*

Sowton M, Haywood P, (1986) National Standards for the Transfer of Digital

Map Data In: *Proc. of the Int. Conf. on the Acquisition, Management and Presentation of Spatial Data* (*Auto-Carto London*), London (UK), September 14–19, 1986, Vol. 1: 298–311

Spooner R, (1990) Advantages and Problems in the Creation and Use of a Topologically Structured Database in: *Photogrammetry and Land Information Systems*, Lausanne (Switzerland), March 6–17, 1989: 73–85

Tassot A, Laurent D, Couesnon T, (1988) Génération d'information topologique sur une base de données géographique (réseaux urbains)

Thorpe J A, Gilbert D M, (1987) AM/FM Requirements of GIS: A Commentary, *Photogrammetric Engineering & Remote Sensing*, **53** (10)

Ulc J, (1989) Intergraph Approach to GIS In: *Photogrammetry and Land Information Systems*, Lausanne (Switzerland), March 6–17, 1989

Ulc J, van der Laan F, (1990) Vector/Raster Integration at Intergraph, *Geo-Informations-Systeme*, **3** (3): 34–43

Veregin H, (1989) Accuracy of Spatial Databases: Annotated Bibliography, Technical Paper 89–9, NCGIA

Veregin H, (1989) A Taxonomy of Error in Spatial Databases, Technical Paper 89–12, NCGIA

Wu S-T, (1989) A New Combinatorial Model for Boundary Representations, *Computer & Graphics*, **13** (4): 477–86

Yost R A, (1988) Can Image Data be Integrated with Structured Data ? In: *Proc. of Fourth Int. Conf. on Data Engineering*, Los Angeles (California, USA), February 1–5, 1988: 602

Chapter 4

CARTOGRAPHIC DATA MANAGEMENT

Peter Haywood
Ordnance Survey, Romsey Road, Maybush, Southampton SO9 4DH, UK

INTRODUCTION

If a number of cartographers formed a committee for the purpose of examining national mapping in any country where digital technology is likely to be used extensively during this decade, with a view to defining what the responsible production agencies should be doing in the year AD2000, what would be said? Would the main objectives be map production or the management of data to support information systems? Would data specifications bear much resemblance to what is seen now? Would the committee recommend some rationalization of existing divisions between collectors and providers of topographic, natural resource, and demographic data (to mention but three types), so that the traditional topographic base for so many other datasets becomes a geographic base for GIS? If so, what are the most appropriate accuracy, resolution, content, and revision criteria for this base?

It is suggested here that this type of exercise is necessary if spatial information systems are to show significant progress by the end of this decade. They will not be held back by technology but might be by inappropriate management of national data resources, which in turn is unlikely to be the consequence of incompetence, ignorance, or lack of determination on the part of individuals but the absence of national, corporate, or professional (cartographic) visions of information system ideals and their implications on data management.

A database management system (DBMS) is a tool for data management, which is a traditional cartographic function in the guise of data collection, processing (say, editing and draughting), and product delivery. Some cartographers have been responsible for maintaining national map series, and the survey records, documents, films, and overlays used for this purpose constituted a database.

It is readily accepted that spatial information systems will have a big impact on cartography in the 1990s, but a question is still being asked: 'What is the significance of GIS?' Many potential users find that what is on offer is distorted by hyperbole, and what is theoretically achievable is not so in practice. Why is this? Is there a technical or data problem?

What is needed is a comprehensive, but succinct, statement of where information systems will ideally lead and what the impact will be, particularly on data management, so that appropriate policies and practices can be adopted. There are technological weaknesses, but these will go away. The outstanding problems will then concern data management, but these cannot be easily solved by innovation. The solutions will take time and will be expensive, and may have to be preceded by redefinition of the business some data providers are actually involved in. We cannot start to overcome data deficiencies until the needs of GIS of the future are defined.

SCOPE OF APPLICABILITY

Every point made in this chapter will not be universally applicable because cartographers and cartographic agencies do a wide variety of things, and expectation is influenced by institutional factors (Chrisman, Chapter 11). For simplicity, the generic term used in this chapter for spatial information system is GIS. This follows the line adopted by some (Rhind and Mounsey 1989) but is in contrast to other opinion which believes that LIS, at least, is sufficiently distinctive to merit its own label and description (Dale 1989). However, comparisons of design and capability between systems and the way in which they are used show that there are significant differences, even though the nomenclature may not be favoured. Some use 'small scale data' and deal mainly with natural phenomena which necessitate overlay and fuzzy techniques. Others use 'large scale data' and deal with man-made (or conceived) things like land parcels where accurate survey (W P Smith 1989) and exact correspondence is necessary (but not necessarily achievable). The reader may find, therefore, that some of the statements made about GIS may be irrelevant to his own use or conception of it.

HISTORICAL BACKGROUND

Theoretically, there should not be fundamental differences between data management needs in a conventional mapping environment and a digital one. Data has to be collected and looked after in both. However, in conventional mapping much that is ideal, such as data integration, is not easily attainable and consequently is not pursued. Computers provide tools which have been missing in the past and therefore raise expectations, introducing new and untested methods to replace those which have been proven and refined over many years.

Two of the biggest conventional mapping problems are (*a*) *keeping maps up to date, and* (*b*) *providing a variety of maps* to meet the needs of specialists. Their cause lies in the production processes and in distribution, which are costly and time consuming. Another problem lies at the user end and is caused by the same inadequate technology. Irrespective of how quickly good quality and updated maps can be supplied, *a replacement map is often an embarrassment to a user* who has put resources into modifying and tailoring the original to his purpose. Hence, technical advances in one of the three main aspects of production — data collection, processing, and distribution — have been difficult and slow to exploit because they simply cause a bottleneck elsewhere and, moreover, the end user cannot benefit easily. Conventional methods change very slowly.

Production flowlines are usually long, with many operations, and measured in years rather than months. There is plenty of time for examination and quality control, and because data are always visible as lists and graphics it is tedious but not otherwise demanding to check and secure. Production monitoring and control methods are facilitated by documents, the movement of which gives a means of monitoring and controlling work flow.

The unit of work in a conventional flowline is usually the 'map sheet' because it is often the only practical way of managing data collection, production, and supply, and the means by which users can cope with their data. Management is easier if an area of interest is regularly divided into sheets, and even a slight divergence from this in order to provide edge overlaps can create difficulty. Production becomes more complicated and data inconsistency can occur. Although it is not unusual now to hear or read that a map is just one method among many of transferring data from collector to user, or that map makers are really in the information business (Haywood 1986, W P Smith 1989), the fact is that maps were not seen this way in the past. Their production was the ultimate objective, and this resulted in distortion and simplification of data. The map sheet constrains much current thinking about data management and to think beyond it has implications on a whole range of database issues.

THE IMPACT OF DIGITAL TECHNOLOGY

Digital technology will have its greatest impact on cartography through its effect on users, that is, those who use maps rather than produce them. Generally, what has been seen so far are digital methods introduced piecemeal into producing agencies as little more than localized technical improvement. In fact, the complications they cause elsewhere in data management can often outweigh the benefits.

From a production point of view, *efficiency gains are now being made as a result of digital technology but until recently there has been little evidence of radically changed production programmes* emerging because different and better things are possible. Although the digital cartographic base and computer technology solve some long standing production problems, it does not necessarily follow that appropriate solutions for the 1990s are already in place.

Users' Requirements

How are users' requirements going to change in the 1990s and dictate what cartographers have to do and produce? From a long list of possibilities three things can be chosen to paint a general picture.

1. The most important is the spatial information system, which will do most to change user attitudes to spatial data. The nature of GIS means that in order to reap maximum benefit the change cannot always be localized, and could mean major re-organisation (De Man 1988, Dangermond 1989). Therefore, expectations for returns will be very high. One of the professed advantages of GIS is that it will allow spatial data to be shared, and if this is the case then *somebody, perhaps remotely connected with a particular GIS, has to take on the responsibility of managing data for a wider purpose than his own.*

2. Specialized systems which could be called subsets of GIS will develop, such as vehicle navigation (Claussen and Mark, Chapter 8) and map access and spatial query through television. These will also be big investments and will be *driven by strictly commercial objectives and up-to-dateness will often be a very important issue*, putting severe demands on data providers.

3. There will be new markets for printed maps fuelled by the ever increasing demand for information in business and pleasure activity where the total need will be large but too localized to merit investment by the users in GIS or digital mapping systems. This market will want *a fast map production service and customization of the output.*

This scenario has a major implication. *The map or data producer will be brought closer to the applications and many of the data management problems dealt with by the large producer will either* (*a*) *have to be faced by the user, or* (*b*) *the producer's problems will be intensified.*

The Production/Management of Maps/Data

Data production and management go hand in hand, and in fact, where topographic or general base maps already exist in graphic or digital form and are regularly updated it becomes more appropriate to talk about management than production. *Data management problems become severe when changing or modifying what already exists*. Update and maintenance are the long term issues, and data capture will imply capture specifically for the purpose of update or enhancement.

The *mechanics* of collection are becoming easier and the real cost is falling as a result of the introduction of a wide range of electronic devices, from hand held keyboards and recorders, to GPS, and sensors in satellites. New techniques are developing on the back of this technology, such as photogrammetric superimposition, which it will be noted, is primarily an update technique (Newby 1990). Things are possible which could not have been contemplated until recently, such as re-adjustment of the complete large scale map base of GB using the digital version backed up by dense control from GPS (Bordley and Christie 1989). Apart from any perceived advantage in doing this it is probably becoming necessary for practical reasons: such is the attainable accuracy of modern survey techniques, very often in absolute rather than relative terms, that fitting surveyed change to older data is difficult and can give the techniques a cost disadvantage instead of the advantage that ought to arise. It is a data management problem which can result in inconsistency and uncertainty about quality. Newby (1990) also reports on the development of a portable interactive edit station (PIES) for field surveyors, again for update, and there are rumours about the impending introduction of a GPS receiver of personal cassette player dimensions (GPS World 1990).

Technology is also changing the subsequent processing of data and product delivery. Although automatic generalization and text positioning are still problematical (Muller, Chapter 1), many manual cartographic processes can be eliminated and maps produced more cheaply. *The speed and lower cost of digital methods mean that customization and currency of maps no longer pose intractable problems* (Rhind and Mounsey 1989, Sowton 1989*a*, 1989*b*, Haywood 1990*a*). Modern printing and communication technologies, in combination with direct digital capture, allow almost instant production of up-to-date maps, and during the 1990s production systems will be installed in high street shops and service stations. Knowledge based systems and automated design techniques (Rase, Chapter 9) will expand their capabilities. Abandonment of some litho printed maps and national map series are now distinct possibilities (Rhind and Mounsey 1989). At this time, the economics of distributed map production may be questionable but most of the technology already exists.

Data collection and map production are ceasing to be technically constrained activities, *although applying new technology is not always straightforward* (Proctor and Newby 1988). The problems that arise are usually associated with data management. While it is possible to theorise about the benefits of working in digital form and using DBMS, *a production manager can legitimately ask how production can be controlled and quality assured in a fast moving environment where the product is not seen until the last minute or not at all except by customers?*. Of course, what is being managed is a database and it is here where control has to be exercised (Haywood 1990*b*), but there is a danger. *If the end products are maps (graphic or digital) the problems and questions will be dealt with as map production issues*, and using new technology will be justified on the basis of lower cost and better maps and mapping services.

The future of cartography is associated with information systems, and these may demand somewhat different data collection, processing, and delivery techniques.

Spatial Information Systems

Despite published 'definitions', use of the term GIS is much abused. Almost any system that displays a digital map or scanned image is now called a GIS (Cowen 1988), rather than described as the up-market digital mapping system which it probably is. Descriptions are necessarily generalized, but also attempt to be all encompassing by including system functions such as data capture, data analysis, display, and output, leaving it open for anyone to place emphasis as seen fit. This leads to difficulty, because *all the technical and data management issues requiring solutions tend to be seen as internal to a particular system* (Chrisman 1990), and hence can be satisfactorily overcome by the right combination of DBMS, hardware, and software within the system. While this view prevails the importance of information systems and what needs to be in place to achieve maximum benefit are not appreciated in the right places.

A significant feature of a GIS is its ability to perform spatial analysis, and this is what sets it apart from CAD, digital mapping, and DBMS (Burrough 1986, Cowen 1988, De Man 1988). It is essentially a decision support system (De Man 1988, Dale 1989, Armstrong and Densham 1990) and not a display or map making system, and if it is to have widespread rather than localized benefit it must bring data together from many different (reliable) sources for the purpose of analysis (Dale 1990). A variation on this latter point is that GIS allows data sharing (Chorley 1988). There is merit in a GIS description which does not attempt to pin down the system components but concentrate on the above issues, as used by W P Smith (1989) for example. However, there is no consensus on these points, and it has been reported that *uncertainty about where GIS stands in relation to the whole field of Information Technology has been inhibiting* (Bracken and Webster 1989).

A categorization of GIS as used and implemented is helpful. Firstly, there are systems which are used mainly for data capture and data management purposes, and it would avoid much confusion if these were called topographic, demographic, or land resource information systems, or simply data capture systems. Secondly, there is the self-contained system with some owner/user maintained data and a number of imported datasets. Significantly, the latter *remain static until eventually replaced in totality or on a map by map basis* by updated versions. Thirdly, and not greatly in evidence at this time, is the dynamic system which regularly imports updates to datasets in order to update the GIS database(s). Fourthly, and more of a concept than a reality but anticipated by some (Smith et al. 1987, Webster 1988), is the national GIS whereby data is imported only when and in the quantity required for analysis by communication network from a number of remote databases. The last three of these categories approximate to those used by McLaren (1990) in his description of GIS information management architecture in which he discusses some of the implications of each.

Unless every GIS owner is going to take on data management problems for datasets *with which he has little expertise* (W P Smith 1989), economic, standards, and common sense arguments dictate that the mega or multipurpose distributed system (Chrisman 1990) is the long term objective, although the dynamic type is an acceptable and possibly necessary interim solution. Chorley (1988) even suggests that data should not be brought together and held centrally because lack of real interest in it will reduce confidence and cause it to become out of date. The potential benefits of distributed systems have been expounded, and so have some of the associated technical and data management problems (Nicholson 1989, Batty 1990), but it is often found that the enthusiastic proponents are system vendors. *Unless cartographers are equally enthusiastic and have a vision beyond technical constraints, the distributed GIS will be difficult to achieve because of data problems.*

At the present time, the self-contained GIS dominates the conceptual view, and this perpetuates the tradition of *compartmentalized data management*. 'Tiles' are used instead of map sheets but these do not avoid all problems (Frank 1988, Chrisman 1990), especially as a method tends to be specific to a system. The 'coverage', or map theme, leads to difficulty because data are collected at different times by different organizations to a variety of standards (Chorley 1988), as amply illustrated by a National Center for Geographic Information and Analysis initiative (Goodchild and Gopal 1989, Goodchild, Chapter 6). The importance of data quality and source information has been well publicized (Chrisman 1986, National Transfer Standards 1987) and although this may assist with software solutions a comment by Burrough is very significant: '...practices in thematic mapping tend to hide natural variation behind a smoke screen of smoothly drawn lines...' (Burrough 1986). There should be a similar cautionary note about disguising inconsistency within data with applied algorithms. *If reaping the full benefit of GIS is dependent upon using different sources for data, it should be asked if a tenuous link based on partitions and coordinated coverages is all that can be expected, or whether something better is needed.* Splitting data into spatial and aspatial components, which takes many forms (Bracken and Webster 1989), allows advantage to be taken of readily available, general purpose database technology for retrieval but it creates an additional data management problem and preserves existing divisions of responsibility between mensuration and classification. Nyerges (1989) states that the successful integration of data from different databases is primarily dependent upon matching of thematic rather than locational descriptions. *If this is so, then one can ask whether the future of GIS depends upon standards for positioning and a topographic base, or more upon standardized descriptions of objects?*

When a GIS is a decision support system, data have to be up-to-date. Data which are out of date, or have been updated by improper procedures, are unreliable (Shand 1989). High value investments and legal matters could be involved, apart from the obvious need to make the system pay on the basis of sound decision making on mundane issues. Existing methods of updating GIS databases are difficult and labour intensive unless coverages or partitions are replaced in their entirety. But surveyed change is only one aspect of a much wider time issue; provisional and planned change is often as vital to decision making as the latest surveyed change, and a record through time and historical geographical analysis sometimes equally so. A continuous record of change can be essential, especially when a GIS is used for cadastre, and snap-shots in time as partitions or coverages cannot handle this adequately (Hunter 1988, Hunter and Williamson 1990). Similar observations can be made about accuracy and quality in general; existing data structures do not allow for adequate recording, access, and analysis of them.

Information systems are perpetuating many conventional cartographic problems. The reasons are pragmatic and probably inescapable at this time. Data structures have to allow for technological weaknesses and procedures for inadequate data, so that system designers and users are being forced to comply rather than innovate. What is less excusable is that other sections of the GIS community are expecting technological solutions and nothing more. During the 1970s it was widely believed that inadequate technology was the stumbling block to progress in digital mapping, but hindsight suggests that the real cause was naivety (Dale 1989). *There is an important distinction between the management of geographical data and its manipulation and analysis* (Batty 1990), and 'data systems are the tools of data managers; information systems are for the decision makers' (Dale 1989). If there is such a distinction then there are also matters which can be dealt with separately. Waiting for technology results in computer systems which replicate mistakes and inefficiencies of the past (Dale 1989), because that part of the solution which comes first dominates the other.

This is especially true when one part of the solution will of necessity take longer to achieve.

The Impact of GIS on Data Management

The right technology for GIS will appear with or without assistance from cartographers, but they have an important contribution to make. The essential point is this: *a new computer innovation may immediately open up new possibilities, but if these also demand changes to data and data management practices it could take many years before that innovation yields any benefit.* Such changes cannot be implemented quickly; all aspects of data management practices and tools have to be examined and modified, including models, structures, security, communications, and update strategies (Coote 1989*a*).

The databases of the future have to cope with a number of data management issues. As data sharing and online access to a number of databases are, to many potential users, the real benefits of GIS and should be the ultimate goals, it is unacceptable to think that the user has to modify data to suit the peculiarities of his own system. *This will degrade accuracy and other quality attributes and mean that data standards cannot exist across systems.* Without these, the validity of information and decisions arising from use of the system will always be open to question. Modifying data is expensive, and as update is clearly an important factor the user will be faced by recurring expense. There is evidence in the UK to suggest that users of digital systems are reluctant to acquire updated topographic data, even though it is available, because of the effort involved in loading it into a database. Here we have a traditional cartographic problem yet to be solved (see Historical Background). Without standards across systems and with update difficulties, users cannot meet some of their objectives.

The search for GIS solutions continues. The seamless database is a possibility (Guptill 1989*a*, Chrisman 1990), but this needs appropriate data structures, exchange formats (Moellering, Chapter 7), and data from providers. Object oriented programming (Newell 1990) and databases (Maguire et al. 1990, Salgé, Chapter 3), and spatial query languages (Robinson, Chapter 5), may be requisites but there are some significant problems (Chrisman 1990), one of which is the 'long transaction' (Newell and Easterfield 1990). Use of the 'object' as a fundamental building block for structures (Gahegan and Roberts 1988) and data integration suggests answers to many questions. Different positional accuracies and quality attributes can be applied to objects (Guptill 1989*b*); there could be time dependent representations of the same object (Hunter 1988, Langran 1989*a*, *b*, Vrana 1989, Hunter and Williamson 1990); and different coverages could refer to the same object without being unduly concerned with accuracy of position and geometry — associated with the basic spatial unit issue (Openshaw 1987, 1990, Chorley 1988). An object could be replaced or a new version created in a database when an update is available (Vrana 1989), perhaps through an exchange format which handles 'change only update' (AM/FM Workshop 1990, Haywood 1990*c*, Moellering, Chapter 7).

It can be assumed that *the management of data about objects instead of maps is a very different business*, and if the hardware and software tools become available they will have little impact unless the 'national database' is put into order (Worrell 1990). While natural phenomena obey few rules and are beyond the wit of man to organize (Burrough 1986), there is much about man-made phenomena and generated data that can be controlled. The collection of statistics about things which should exhibit meaningful relationships by different non-aggregatable spatial units (Brusegard and Menger 1989) and allowing the boundaries of the units to be changed at whim, suggest that national spatial databases are not in line with Information Technology ideals.

Technology already available indicates that it is not impossible to improve data management methods, and some agencies have shown a willingness to look ahead. In France (Salgé and Piquet-Pelorce 1986, Salgé and Schlafer 1989), the USA (Guptill 1990), the UK (Coote 1989*a*, Haywood 1989, Sowton 1989*b*), and Germany (Barwinski and Bruggemann 1986) there have been research projects around the themes of topographic databases and topographic information systems with objectives of better data management and data for spatial information systems. In each case, every effort has been made to look beyond map production programmes and technical constraints, to the extent that at IGN (Paris) it has yet to be demonstrated that maps can be made from the prototype database, and at Ordnance Survey (UK) a relational DBMS and futuristic data structure have been used in the full knowledge that they would be inefficient with present day hardware and software (N S Smith 1987). However, as there is a limit to how far these agencies are allowed to speculate, the problems are immense (Pornon and Gateaud 1990), and since data integration and unilateral action are incompatible, they are not going to solve them unaided.

The reason why no data producing agency, public or private sector, will make radical changes to its data and data management practices is uncertainty about the future requirement. *Even small changes can involve enormous effort and expense.* Ordnance Survey (UK) has started to edge match its large scale digital maps, of which there will be nearly 220 000 by 1996, and the first part of the edge match programme will complete only a quarter of these in three years. For it to be worthwhile, the whole system of revision has to be modified so that matched edges are maintained on update (Coote 1989*b*, Sowton 1989*b*). The fundamental requirement is that field and photogrammetric survey systems handle apparently seamless data, but in support of this internal storage and transfer, methods also have to be changed to enable dating and flagging (data about data) which becomes essential data management practice once the map sheet ceases to be the work unit. In a complete digital system like this one, corruption is a very real danger and new validation and security procedures have to be adopted (Coote 1989*a*). The magnitude of production control and management problems takes on new dimensions (Haywood 1990*b*); in simplistic terms, it can be said that edge match requires eight digital maps to be made available and worked on at a workstation instead of one, so eight maps are no longer available for work elsewhere. Ordnance Survey would like to justify this on the grounds that it is preparing for the information systems of the future; in reality, it has had to demonstrate that there are compensatory benefits in its more traditional cartographic activities (Sowton 1989*b*). If it is to go further, say introduce and work with objects, *it may find that information system requirements are more expensive to meet* and there are no other compensatory benefits, nor reliable evidence to show that this is a worthwhile investment.

There is a Catch 22 situation. *Information systems are being designed (in part) on the assumption that available data will be imperfect. This prohibits effective use of systems, particularly in respect of data sharing and update. Alternative technological solutions are proposed and researched, which in turn disguises the true data need, and until this is apparent the right data will not be made available.*

PRACTITIONER'S SUMMARY

There have been significant gains in using digital technology for map production since 1985, but prior to this development was slow. Digital methods compared unfavourably with conventional ones in terms of cost, and there were few gains from better products to compensate. Two important goals for mapping were more up-to-date and customized maps, but the introduction of expensive and limited technology into selected parts of flowlines merely chipped away at costs and caused consequential problems elsewhere without stimulating a drive towards these.

There was a technological breakthrough in the 1980s. It was not specific to a device or method but was a host of developments in photogrammetry, field surveying, and map production which rode on the back of miniaturization and falling costs in computer hardware and satellite technology. Production systems have since been developed which are totally digital — from data capture to product delivery — instead of being computer assisted, and it is possible to deliver an updated map within days of survey being completed. To some extent it is also possible to customize the product within the same timeframe. The conversion of data into graphics can now be distributed so that there is more interaction between producer and customer and large stocks of printed maps are no longer essential.

Map production is becoming 'black box' technology to the map user, but this unfairly reflects the pressure that this technology has put on producing agencies. The idea that data can consist of small free standing packages — maps in digital form — can no longer prevail, but this has been a fundamental plank of data management, production control, and marketing policies for many years. Specifications have needed questioning because customization and the new markets opened up by it require flexibility in data. Traditional ways have been changed, but one of the most important indicators of achievement over the last twenty years is that higher level management now recognizes that in order to reap maximum benefit from digital technology there have to be changes in policy, organization, and attitude.

Changes to data and method do not happen quickly and without cost, and usually it is necessary to implement both at the same time. A possible blot on the cartographic achievements of recent years is that the technological breakthrough had to come before there was any vision of what lay beyond it. Now there is talk of re-adjusting topographic bases, dispensing with national map series, and only producing what customers want when they want it, but the quality of databases lags behind technical ability to do these things. Had there been a vision it may have been possible to implement improvement to data much earlier. This may have saved costs — reworking to new specifications has high costs — but more significantly it would have reduced delay. The scientific and professional community should take note in respect of spatial information systems.

PROJECTIVE SUMMARY

The next major technological breakthrough will concern spatial information systems, but as these have had very little impact on data producing agencies and national databases so far it is not out of order to ask whether inappropriate data management will prevent desirable and technically possible goals from being achieved? There is good GIS technology and it is certain to get better. There is plenty of data already available, but are they ideal for use in GIS? What is not available is good GIS theory which could be the basis for policy and direction.

In any country or region there is usually only one source of good data of a particular type, whether it relates to topography, geology, population statistics, or property. Good data implies that standards have been met, certain skills applied in its acquisition, and its ongoing availability is backed up with a maintenance programme. GIS requires data to be managed with integration in mind and not alternative data sources because technology makes capture easy. Error models are essential to ensure quality rather than make allowances for poor quality during analysis.

It is an expensive, slow process to change a production system, and frequent unplanned changes resulting from a technology chase cannot be tolerated. From a production manager's point of view, this is exactly what is happening with GIS, and it is impossible to respond to every new innovation. Production programmes and their supporting systems have to have long term objectives and a plan of campaign.

More needs to be done about data management, and it is the biggest issue facing GIS and cartographers. While this is apparent to some, as many of the references used in this chapter indicate, it is clouded by contradiction, misrepresentation, and vested interest. Bringing 'the national database' into line with information system requirements is a tall order and will need policy decisions at high levels, but policy makers will be tempted to ask 'if the mechanics of mensuration and data recording are becoming so simple, and GIS is the panacea for spatial analysis and data management, where lies the problem?'

The long term problem is that the technology could arrive and data will be in a mess. The short term problem is that there is no apparent consensus of opinion amongst cartographers, as a professional group, indicating this and identifying remedial action. It is not the purpose of this chapter to identify technological research nor research into data management techniques, but research to establish a view of the future from which ideal data management goals can be established. Data management is the responsibility of line managers in cartographic/data production units, but as the interfaces with the outside world become wider and more complex with the introduction of spatial information systems, managers will need direction and guidance from the cartographic profession as a whole. The alternative is to let them deal individually with users and system suppliers, and if this is allowed to happen it will work against the multipurpose, on-line GIS.

A Vision for GIS

Most of what needs to be said has probably already been published somewhere, but it is scattered, contradictory, and hidden amongst irrelevances. A professional group should bring the essentials together and describe the GIS of the future. A model is needed, as free of perceived technological, traditional cartographic concepts, and institutional barriers as possible. Long, medium, and short term models may be necessary, based upon advice from the computer industry and research communities as to when some essential technology will be available so that timescales can be developed.

A Geographic Base

From the GIS model(s), a geographic base should be described, as suggested in the opening paragraph to this chapter. Again, it should be a reasoned model, with suggestions about accuracy, quality, content, temporal data, and so on, but not a specification. It may bear little resemblance to topographic bases as seen now and cut across many of the boundaries that have been erected between data collecting agencies.

Contributing Databases

Spatial information systems will need a number of databases in addition to the fundamental base. What are these in light of the conclusions from the first two agenda items above?

CONCLUSION

A research agenda such as this is a pre-requisite to the resolution of data management issues. Unless a source of reference produced by a reliable, respected, and international professional group is made available within the next two or three years, the 1990s could be remembered for endless questions about the nature of GIS and arguments about who should do what. It is widely recognized that hardware and software capabilities are expanding faster than human abilities to apply them. One aspect of this is that data management, which is more of a human than technological problem, is already lagging behind capabilities to use data effectively. It will not catch up if technology is chased; it has to be anticipated. Cartographers should not wait for Release 2.0 of system 'GIS XYZ' to find out where they go next.

REFERENCES

The references are of two types. Firstly, to give a lead into technical issues; most of those selected are recently published works containing further references which allow back-tracking on a particular subject. Secondly, to support the opinion expressed or alluded to, particularly from a practitioners point of view. The two types are designed to be complementary and strike a balance between the positiveness and (often) optimism of those whose business is the development and provision of technology, systems, and software, and the questioning, uncertainty, and (often) pessimism of users and data providers.

AM/FM (1990) Proceedings of a workshop *Towards a Common International Transfer Format*. Montreux, AM/FM Europe, October 1990

Armstrong M P, Densham P J (1990) Database organisation strategies for spatial decision support systems. *International Journal of Geographical Information Systems* 4 (1): 3–20

Barwinski K, Bruggemann H (1986) Development of digital cadastral and topographic maps — requirements, goals, and basic concepts. In: Blakemore M (ed.) *Proceedings of Auto Carto London* **2**: 76–85

Batty P (1990) Exploiting relational technology in a GIS. *Mapping Awareness* **4** (6): 25–32

Bordley R F, Christie R R (1989) Ordnance Survey data transformation procedures. In: *Proceedings of a seminar at Nottingham University*, UK, December 1989

Bracken I, Webster C (1989) Towards a typology of geographical information systems. *International Journal of Geographical Information Systems* **3** (2): 137–52

Brusegard D, Menger G (1989) Real data and real problems: Dealing with large spatial databases. In: Goodchild M, Gopal S (ed.) *Accuracy of spatial databases*. New York, Taylor & Francis: 177–85

Burrough P A (1986) *Principles of Geographic Information Systems for land resources assessment*. Oxford, UK, Oxford University Press

Chorley R (1988) Some reflections on the handling of geographical information. *International Journal of Geographical Information Systems* **2** (1): 3–9

Chrisman N R (1986) Obtaining information on quality of digital data. In: Blakemore M (ed.) *Proceedings of Auto Carto London*, 1986.

Chrisman N R (1990) Deficiencies of sheets and tiles: Building sheetless databases. *International Journal of Geographical Information Systems* **4** (2): 157–67

Coote A M (1989*a*) Managing a Large Spatial Archive. In: *Proceedings of AGI* 89 (Association of Geographic Information Annual Conference), Birmingham, UK, October 1989: B.3.1–B.3.5

Coote A M (1989*b*) Current developments in field-based digital mapping systems at Ordnance Survey. In: Shand P J, Moore R V (ed.) *The Association for Geographic Information Yearbook* 1989. London, Taylor & Francis: 113–19

Cowen D J (1988) GIS versus CAD versus DBMS: What are the differences? *Photogrammetric Engineering and Remote Sensing* **54** (11): 1551–5

Dale P F (1989) Land information systems and the land parcel. In: Shand P J, Moore R V (ed.) *The Association for Geographic Information Yearbook* 1989. London, Taylor & Francis: 59–63

Dale P F (1990) Geographical information systems. *Government Computing* **4** (4): 25–7

Dangermond J (1989) The organisational impact of GIS technology, *ARC News*. ESRI, Summer 1989: 25–6

De Man W H E (1988) Establishing a geographic information system in relation to its use: A process of strategic choices. *International Journal of Geographical Information Systems* **2** (3): 245–61

Frank A (1988) Requirements for a database management system for a GIS. *Photogrammetric Engineering and Remote Sensing* **54** (11): 1557–64

Gahegan M N, Roberts S A (1988) An intelligent object-oriented geographical information system. *International Journal of Geographical Information Systems* **2** (2): 101–10

Goodchild M, Gopal S (eds) (1989) *Accuracy of Spatial Databases*. New York, Taylor & Francis

GPS World (1990). Eugene, Oregon, USA, GPS World Corporation, March/April 1990

Guptill S G (1989*a*) Speculations on seamless, scaleless cartographic databases. In: *Proceedings of AutoCarto* 9, Baltimore, April 1989: 436–41

Guptill S G (1989*b*) Inclusion of accuracy data in a feature based, object oriented data model. In: Goodchild M, Gopal S (ed.) *Accuracy of Spatial Databases*. New York, Taylor & Francis: 91–7

Guptill S G (ed.) (1990) An enhanced digital line graph design. *US Geological Survey Circular* No 1048, Reston, Virginia, USGS

Haywood P E (1986) The differing requirements of producers and users of data and the design of topographic/cartographic databases. *Nachrichten aus dem Karten-und Vermessungswesen* (Institut fur Angewandte Geodasie) Series ll (44): 57–63

Haywood P E (1989) Structured topographic data: The key to GIS. In: *Proceedings of AGI* 89 (Association of Geographic Information Annual Conference), Birmingham, UK, October 1989: B.1.1–B.1.4

Haywood P E (1990*a*) La production cartographique de l'Ordnance Survey en 1990. *Bulletin du Comité Français de Cartographie* September 1990 No. 125: 42–7

Haywood P E (1990*b*) An integrated production and business strategy. In: *Proceedings of Euro Carto V*11, Mallorca, Spain, April 1990

Haywood P E (1990*c*) The case for NTF (National Transfer Format). In: Proceedings AM/FM workshop *Towards a Common International Transfer Format*. Montreux, October 1990

Hunter G J (1988) Non-current data and geographical analysis: A case for data retention. *International Journal of Geographical Information Systems* **2** (3): 281–6

Hunter G J, Williamson I P (1990) The development of a historical digital cadastral database. *International Journal of Geographical Information Systems* **4** (2): 169–79

Langran G (1989*a*) Accessing spatiotemporal data in a temporal GIS. In: *Proceedings of AutoCarto* 9, Baltimore, April 1989: 191–7

Langran G (1989*b*) A review of temporal database research and its use in GIS applications. *International Journal of Geographical Information Systems* **3** (3): 215–32

Maguire D J, Worboys M F, Hearnshaw H M (1990) An introduction to object-oriented GIS. *Mapping Awareness* **4** (2): 36–9

McLaren R A (1990) Establishing a corporate GIS from corporate datasets — the database issues. *Mapping Awareness* **4** (2): 52–8

National Transfer Standards (1987, 1989). London, Association for Geographic Information, Royal Institution of Chartered Surveyors

Newby P R T (1990) Digital map revision at the Ordnance Survey. *International Archives of Photogrammetry and Remote Sensing* Vol. 28 part 4: 146–54

Newell R G (1990) Who cares about object orientation? *Mapping Awareness* **4** (4): 10–11

Newell R G, Easterfield M (1990) Version management — the problem of the long transaction. *Mapping Awareness* **4** (2): 43–7

Nicholson R L (1989) LANS, WANS, and VANS: Networking as a component of spatial information systems. In: Shand P J, Moore R V (ed.) *The Association for Geographic Information Yearbook* 1989. London, Taylor & Francis: 285–9

Nyerges T L (1989) Schema integration analysis for the development of GIS databases. *International Journal of Geographical Information Systems* **3** (2): 153–83

Openshaw S (1987) *Spatial units and locational referencing*. Department of the Environment

Openshaw S (1990) Spatial referencing for the user in the 1990s. *Mapping Awareness* **4** (2): 24–9

Pornon H, Gateaud J (1990) Why a consistent method in the implementation of a GIS is necessary. In: *Proceedings of EuroCarto V111*, Mallorca, Spain, April 1990

Proctor D W, Newby P R T (1988) Revision of large scale maps at the Ordnance Survey. *International Archives of Photogrammetry and Remote Sensing* **27** (B4): 298–307

Rhind D, Mounsey H (1989) GIS/LIS in Britain in 1988. In: Shand P J, Moore R V (ed.) *The Association for Geographic Information Yearbook* 1989. London, Taylor & Francis: 65–74

Salgé F, Piquet-Pelorce D (1986) The IGN small scale geographical database (1: 100 000 to 1: 500 000). In: Blakemore M (ed.) *Proceedings of Auto Carto London* **1**: 433–46

Salgé F, Schlafer M N (1989) A geographic data model based on HBDS concepts: The IGN cartographic database model. In: *Proceedings of Auto Carto* 9, Baltimore, April 1989: 110–17

Shand P J (1989) One way ahead. In: Shand P J, Moore R V (ed.) *The Association for Geographic Information Yearbook* 1989.London, Taylor & Francis: 155–8

Smith N S (1987) Data models and data structures for Ordnance Survey. In: *Proceedings of SORSA Symposium*, April 1987, Durham, England

Smith T R, Menson S, Star J L, Estes J E (1987) Requirements and principles for the implemention and construction of large scale geographical information systems. *International Journal of Geographical Information Systems* **1** (1): 13–31

Smith W P (1989) Information society: Fact or fiction? In: Shand P J, Moore R V (ed.) *The Association for Geographic Information Yearbook* 1989. London, Taylor & Francis: 51–7

Sowton M (1989*a*) Ordnance Survey and the future. In: Shand P J, Moore R V (ed.) *The Association for Geographic Information Yearbook* 1989. London, Taylor & Francis: 103–12

Sowton M (1989*b*) Ordnance Survey — Topographic Database Project, DFUS, and Project 88: One year on. In: Shand P J, Moore R V (ed.) *The Association for Geographic Information Yearbook* 1989. London, Taylor Francis: 121–32

Vrana R (1989) Historical data as an explicit component of land information systems. *International Journal of Geographical Information Systems* **3** (1): 33–49

Webster C (1988) Disaggregated GIS architecture: Lessons from recent developments in multi-site database management systems. *International Journal of Geographical Information Systems* **2** (1): 67–79

Worrell L (1990) Improving the quality of database UK. *Mapping Awareness* **4** (3): 31–6

Chapter 5

SPATIAL QUERY LANGUAGES

Vincent B. Robinson
Institute for Land Information Management, University of Toronto, Mississauga, Ontario, Canada L5L 1C6

INTRODUCTION

A query language (QL) is a high-level computer language for the retrieval and modification of data held in databases or files (Samet 1981). It is usually interactive, online, and able to support queries that are not predefined. Users of QLs are usually assumed to have limited technical expertise. Therefore, user interaction tends to be of limited complexity. Many different QL interfaces can be built on top of vendor-supplied QL (Jarke and Vassiliou 1985). Thus, the basic rationale for a query language is to allow users to manipulate data without burdening that user with details of data structures and storage.

The term *query language* is a misnomer. The English verb *query* suggests retrieval only, whereas query languages typically provide update, delete, and insertion operations as well (Date 1983). In the case of a geographic information system (GIS), a geographic database is a collection of sharable geographic data encoded in various formats. A geographic database system (GDBS) provides an integrated collection of geographic data for access by a large number of users. The GDBS is the core of a geographic information system (GIS). It may, or may not, contain relationships between geographic entities in explicit form. SpQL is the primary avenue through which a user interfaces with a GIS. As such it fulfils several roles. One of the primary roles is to allow user access to a GIS database through retrieval. As geographic databases have expanded the importance of browsing as a component of retrieval has grown. SpQLs have traditionally been the mechanism through which users have been able to compose maps. Thus, an important role of SpQL has been to

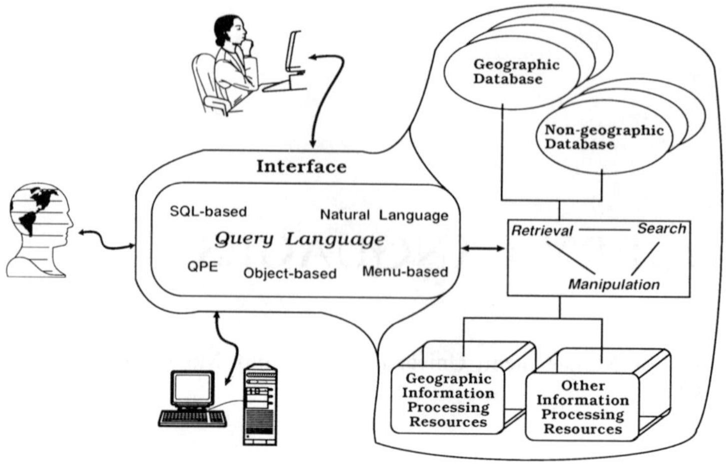

FIGURE 1. Conceptual relationships among GIS, Query Languages, interface, and GIS users

express analytical needs of users such that geographic analysis can be accomplished. Thus, SpQLs are but a piece of the more general interface which a GIS user must negotiate in order to browse, retrieve, analyse, or map (Figure 1).

TRENDS AND DEVELOPMENTS IN SPATIAL QUERY LANGUAGES

A review of existing QLs (Vassiliou and Jarke 1984) suggested that their development stems from the need for simple end user interfaces and theoretical conceptions of programming language or database research. As SpQLs are a subset of QL in general, the development of SpQLs stems from similar imperatives. Likewise, development of SpQLs have been influenced in similar ways by the same factors that have governed development of QLs. Before considering some of the major avenues of development in SpQL let us consider trends and developments in QL.

Development of QLs can be illustrated by relating the functional capabilities of a system and its usability. Roughly speaking, functional capabilities refer to what one can do with the system, whereas usability refers to the effort of actually doing it. Figure 2 summarizes Jarke and Vassiliou's (1985) analysis of QL development trends. One group of QL developers, originating from the disciplines of programming languages and database theory, concentrates on the syntactic form and semantic meaning of database interactions. Languages developed by these disciplines are characterized by the full specification of an operation by a command or a sequence of commands. Starting from formal, mathematically oriented language concepts, this group has moved to *English-like* keyword languages and then to restricted natural languages. The

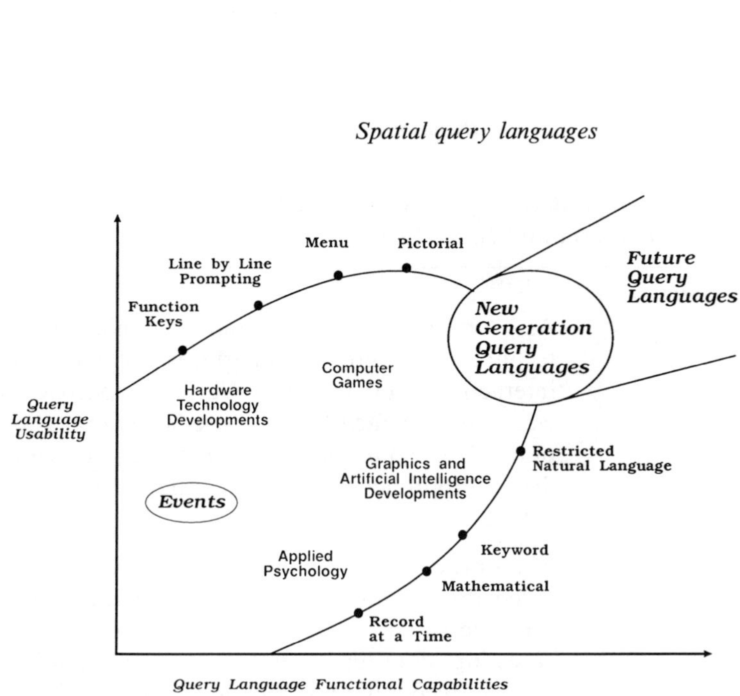

FIGURE 2. Query Language trends and their relationship to events (Source: Jarke and Vassiliou 1985)

overall trend has been towards more *user friendliness* while preserving general functional capabilities. A second group started from the ergonomic analysis of the interaction of computer-naive end users with computer systems. Whereas simple systems use function keys or line-by-line prompting, more complex systems involve the use of menu selection or graphical interaction with the database. These represent a trend towards more functional capabilities while remaining novice-oriented.

The two streams evolved fairly independently with one serving the more sophisticated user with the other serving the less sophisticated user. The languages that have been developed are referred to as *previous generation* QLs. More recent developments are leading to an overlap of the usage area for both groups. The challenge, therefore, is to integrate both approaches into functionally powerful query languages for relatively unsophisticated users, the *new generation* QLs. Emergence of new generation QLs has followed advances in several related areas such as Hardware Technology Developments, Graphics and AI Developments, Applied Psychology, and Computer Games. The aspects of these areas influencing QL developments are summarized in Table 1.

SpQLs have progressed since the early days of automated cartography. In the early days, command-driven packages such as SYMAP required users to adhere to a rigid, fixed-format which was not all that far removed from the development language. One of the early attempts at combining interactive microcomputer technology with a menu-driven geoprocessing system is described by Robinson (1980) and Robinson and Coiner (1986). Figure 3 shows a small portion of the menu scheme of the Urban Datax Management System (UDMS) (Robinson 1980, Robinson and Coiner

TABLE 1 Some Relationships Between Advancing Fields and Query Language Development (Source: Vassiliou and Jarke 1984)

Hardware Technology Developments	As microcomputer and minicomputer technology progressed, interfaces became more interactive and friendly. Various approaches were used to facilitate the user's manipulation of a GIS. Some early systems used a menu-driven approach that is, in some form, still with us. Microprocessor technology and devices such as video disks, and optical storage devices, allow for increased capabilities in storing and accessing data in several forms. In addition, voice recognition/synthesis and pointing devices are leading to increased use of multimedia interactions with geographic databases.
Computer Graphics and Artificial Intelligence Developments	The ability to display information in the most natural and dense form, that of an image, coupled with high-resolution display devices and use of color greatly contribute to the immediate comprehension of query output and the direct representation and manipulation of objects of interest.
Applied Psychology	There is interest in bridging the gap between researchers concerned with the human factors aspects of query languages and their colleagues who are primarily concerned with the technology of QL design (Reisner 1981). Because of the tremendous possibilities for interaction with new generation QLs, the need for scientific methods to deal with the complex considerations of convenience, friendliness, and effectiveness of QLs has become apparent in psychological studies of QLs (e.g. Small and Weldon 1983; Welty and Stemple 1981).
Computer Games	The pleasant and simple interaction of computer games has revolutionized the way in

TABLE 1 (*Continued*)

	which people perceive interfaces to computer systems and helped remove many psychological barriers that people have when faced with a computer system. The success of computer games has influenced QL designers (Malone 1982).

1986) which should illustrate how menu systems can be a cognitive burden to users, especially users that have become familiar with a particular piece of software.

In addition to menu-based systems there are now pictorial query languages (PQL), query-by-pictorial-example (QPE), extensions to the structured query language (SQL), and natural language (NL) query processing systems. Many of these developments have a linkage to the heritage of relational database systems. The relational approach to pictorial, or geographical, database systems was first proposed in the 1970s (eg Mantey and Carlson (1979)). Chang and Kunii (1981) consider a GIS as an example of a relational database approach to pictorial database management. In the past many papers have suggested using the relational database as the underlying structure for a GIS. The structural power of a relational database requires data aggregated by and associated with geographic units. In Chang et al. (1979) the image is called a map and the pictorial relational database is called the d-map. The process of converting the map into a d-map is called abstraction. The inverse of abstraction is materialization. Extensive results for integrated relational database design led to the conclusion that the traditional relational algebra is insufficient to manipulate tabular, graphical, image, and geographical data. For example, traditional relational algebra could not analyse spatial relationships among various geographic objects. In pictorial database systems a set of picture operations called picture algebra was designed for storage, retrieval, manipulation, and transformation of spatial data. Geographical data includes a typical case in which the number of different relations is almost equal to the number of data instances (Chang and Kunii 1981). Thus, the relational model originally proposed by Codd (1970) is not convenient.

Query by Pictorial Example (QPE)

Chang and Fu (1980, 1981) proposed query-by-pictorial-example (QPE) as a picture query language for IMAID an integrated relational database system interfaced with an image analysis system. Most queries are specified from the relations in Table 2 that are parts of a database of

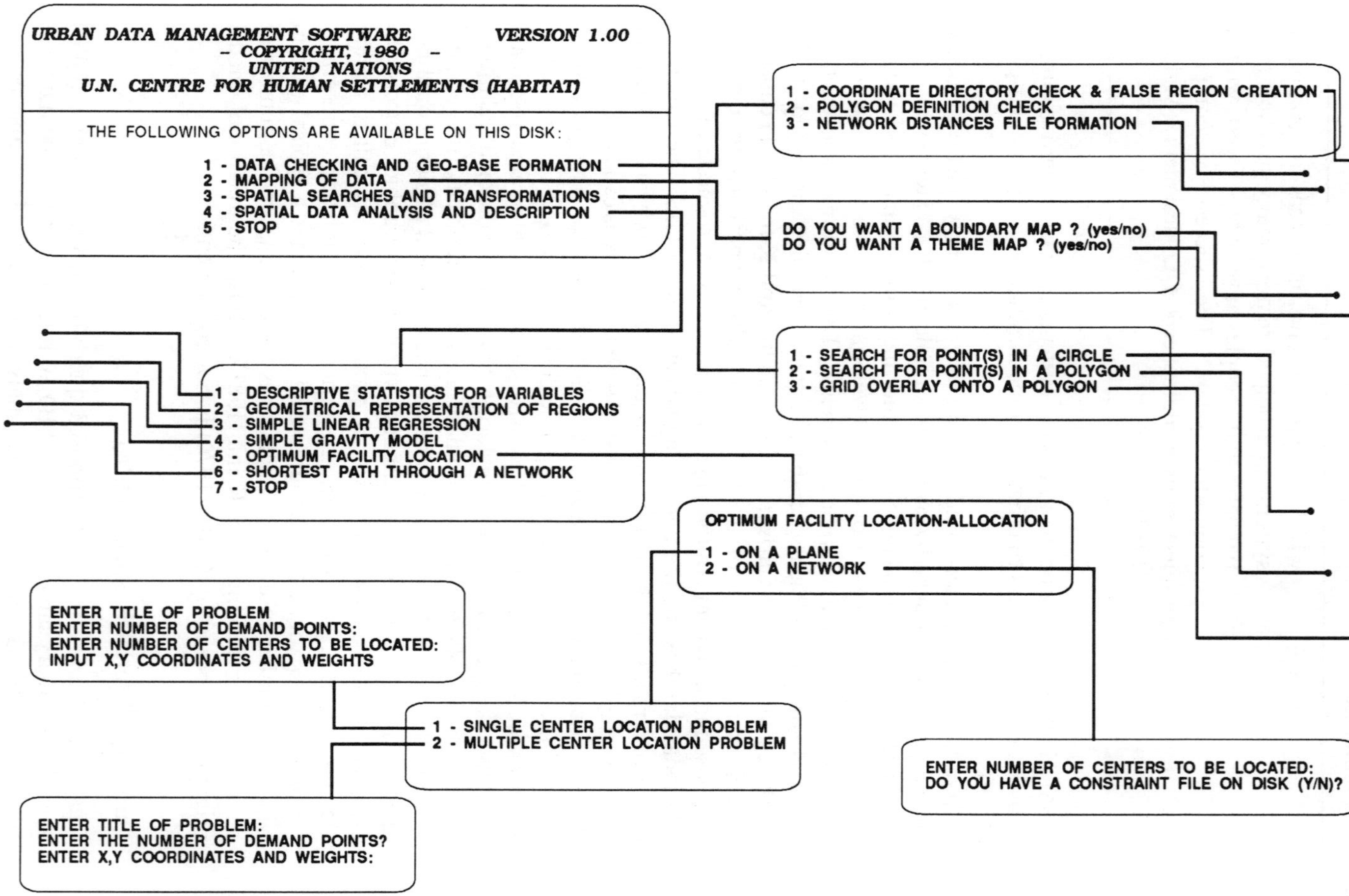

FIGURE 3. Diagrammatic depiction of a the menu-based Query Language of the Urban Data Management Software package (Source: Robinson 1980)

Landsat images and digitized maps. Processing sets are used to recognize boundaries of regions and connected line segments on digitized pictures. This QPE-based system included six classes of manipulation capabilities — conventional manipulation, pictorial entity construction, pictorial attribute manipulation, image-sketch-relation conversion, pictorial example, and similarity manipulation.

Another high level query language, PICQUERY, has been designed to reside as a software layer above a pictorial database management system (PDBMS). It is intended to be the interface through which the user may access conventional relational databases using QBE and at the same time pictorial databases using PICQUERY. In concept, PICQUERY and QBE are seen as one single language by the user (Joseph and Cardenas 1988). PICQUERY is proposed by Joseph and Cardenas (1988) as a new tabular query language based on QPE (Figure 4). It is claimed that PICQUERY has features of both a menu-driven query language interface as well as that of a tabular query language. The PICQUERY language commands may operate on the whole pictorial database or a set of picture-object identifiers. A picture is a distinctly identified, independent image stored in the pictorial database. An object is a named region, line, point, or partial picture with which clearly defined features are associated. It is built on top of a grid-based PICDMS. However, vector-structured formats may be handled if they are first converted to the grid-structure of PICDMS. Since there are a large number of operations, the user will not be able to remember all the operations at any one time. Hence, PICQUERY initially conveys to the user a list of the available operations. If the user types in OPERATIONS P. the system will print all legal available operations with a brief explanation.

Structured Query Language (SQL) Developments

The widespread use of SQL as a database management system language led to the ANSI (1986) standard. This situation has had a profound effect upon how GISs are being developed. Many commercial GISs incorporate to some extent a RDBMS or linkages thereto. Perhaps of more importance is the fact that organizations are beginning to consider GIS implementation to be an integral part of their existing corporate information resources development strategy. This latter consideration provides a compelling incentive for the use of SQL by GIS.

There have been several GIS-related developments using SQL or SQL-like languages (Frank 1982, Abel and Smith, 1986, Waugh and Healey, 1986, Herring et al. 1988, Charlwood *et al.* 1987, Ingram and Phillips, 1987, Keating et al. 1987, Roussopoulos et al. 1988, Westwood and Brinkman, 1988, Goh 1989, Tanaka and Ikeda 1989). A review (Robinson 1990*a*) of SQL and GIS suggested several general issues that have been addressed by these developments. A common characteristic of the above

TABLE 2 Relations for Query-by-Pictorial-Example Specification (Source: Cang and Junii 1979)

ROADS(FRAME,ROID,X1,Y1,X2,Y2)
RONAME(FRAME,ROID,NAME)
POS(FRAME,XSIZE,YSIZE,XCEN,YCEN,LOC)
CITIES(FRAME,CHD,X1,Y1,X2,Y2)
CINAME(FRAME,CHD,NAME)

The ROADS and CITIES relations are defined on the same attributes that specify the frame number, identification number, and x–y coordinates of the two endpoints of each line segment.

The ROADS relation is constructed from Landsat.
The CITIES relation is constructed from digitized maps.

SQL efforts was a means of defining spatial context. The window definition operations are a primitive form of defining the bounds of spatial query and operations. This idea has been incorporated into some systems as a spatial database from which projects are drawn for subsequent processing. Some define the spatial context outside SQL then use SQL to operate upon the delimited database. Typical QLs when applied to GISs are lacking in the ability to handle and produce graphics. Thus, another common extension driven by the demands of GISs is the incorporation of graphics into SQLs used on geographic databases.

Extending SQL to deal with geographical queries has led to the specification of classes of spatial operators. Terms used to denote spatial operators are similar across some of the above implementations. Robinson (1990*a*) noted that there appears to be some convergence towards a common vocabulary that might serve as a useful basis for developing standards for defining spatial extensions to SQL. However, similarity of terms does not ensure that the operations are equivalent. One of the fundamental problems in determining equivalence is the difference in data models upon which the operators are based.

Object-based Developments

Upon some introspection, it would seem that users naturally wish to retrieve and manipulate spatial objects. Thus, there have been some efforts to develop systems that allow users to retrieve and manipulate geographic objects (Herring 1987, Manola et al. 1987, Herring et al. 1988, Mohan and Kashyap 1988, Orenstein and Manola 1988, Menon and Smith 1989).

PROBE is a research project into object-oriented database systems that is to provide a general purpose database system for applications involving spatial and temporal data and other kinds of data with complex structure

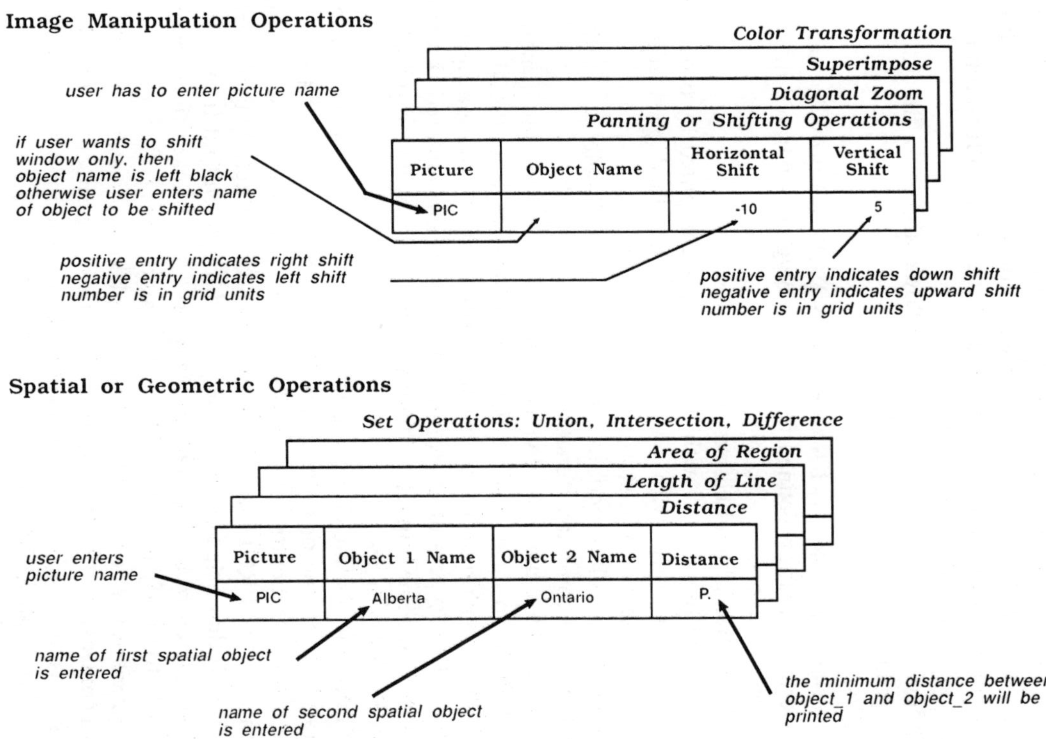

FIGURE 4. Examples of PICQUERY operations as QPE queries (Source: Joseph and Cardenas 1988)

(Orenstein and Manola 1988, Manola et al. 1987). The PROBE data model (PDM) has two basic types of data objects — entities and functions. An entity is a data object that denotes some individual thing. Entities with similar characteristics are grouped into collections called entity types. For example, a database might have an entity type feature representing geographic features. Properties of entities, relationships, and operations are all represented in PDM as functions. Thus, in order to access properties of an entity or other entities related to an entity or to perform operations on an entity, one must evaluate a function having the entity as an argument such as:

population(CITY) → INTEGER allows access to the value of the population attribute of a CITY entity

and the function

overlay (LAYER − 1, LAYER − 2) → LAYER3 provides access to an overlay operation defined for sets of polygons separated into different coverage layers.

There are two general classes of functions:

computed functions—with output values computed by procedures

stored functions—with output values determined by a conventional database search of a stored function extent

References to all functions are treated syntactically as if they were references to computed functions.

Entity types may be divided into subtypes forming generalization hierarchies. In an extensible database system such as PROBE it is not possible to build into the query processor's implementation the definition of each data type and operation that may be encountered. Extensibility implies that some important details of the system will be provided in user-defined object types.

Many queries, including spatial queries, can be expressed in terms of iteration over one or more collections of objects, and the application of one or more functions to each object or to a group of objects within the iteration. The PROBE approach is to provide a generally useful geometry filter(GF) that helps optimize such nested loops. Output from GF is a set of candidate objects that satisfy the query. An object in the candidate set is likely to satisfy the query. The set of candidates is then refined to yield the precise answer by applying user-supplied predicates (eg distance).

In contrast to PROBE, Herring (1987) and Herring et al. (1988) discuss an extension of SQL based on an object-oriented model rather than the relational model. They suggest that the objects can be easily mapped to the relational environment via the entity-attribute-relation data model. A limited number of spatial operators are defined by Herring et al. (1988). The operators are organized as select, derivation, and functional attribute type operators. A multiquery is a sequence of queries each of which uses or modifies results of the preceding query. The concept of multiquery is exploited in what they refer to as the macro expander process. These macros allow users to extend the language by defining groups of operations or tokens as a single unit. The macro facility combined with the capability of accessing procedural code from a query means that user-defined extensions and modifications can be accomplished. Herring et al. (1988) illustrate how this derivation operation can be accomplished in their extension to SQL. This approach uses the macro facility to construct two macros which can be used in a multiquery sequence to answer a query that requires spatial joins subject to attribute constraints. Such a query is a common requirement of suitability analyses.

Natural Language

Natural language (NL) can play many roles in GIS (Robinson et al. 1985). Most NL issues in GIS eventually influence the SpQL. It has been argued that since most users are competent in using natural language, it would be the ideal language for database interaction. The implication being that training would be reduced, or eliminated, and users would have no inhibition to using the computer (Schneiderman 1978). Thus, natural language query systems are an attempt to increase the naturalness of user interaction with GIS. Kasturi et al. (1989) describe a typical natural language front end (NLF) to their map data processing system that converts NL input to the rigid syntactic form required by their query processor. The NLF consists of a preprocessor and a top-down parser. The preprocessor verifies that each word in the query is a part of its lexicon. It interacts with the user to correct any spelling errors and to identify action phrases not a part of its vocabulary. Queries are parsed from left to right. The parser attempts to match its semantic grammar to the input query. It begins with the topmost rewrite rule and, using a semantic-based transition network, moves recursively towards lower, more specific rewrite rules.

Based on rewrite rules, the Action and Display-Object nonterminals form a template. This query represents the main query of the NL input. If there are no Qualifiers, the parsing is complete and the corresponding output is generated. Otherwise the processor notes the need for additional parsing by placing an asterisk at the end of the current output. The process then recursively parses the remaining part of the input query. Output from NLF is input to the query processor which then generates the appropriate calls to the image processor.

Pereira et al. (1982) describe a computer system, interrogatable in a flexible subset of Portuguese which embodies and assimilates expert knowledge on environmental biophysical resource evaluation. It is capable of explaining the application of that knowledge to a territorial database and answering questions regarding its linguistic abilities. Queries can be in natural language or may result from following the options in a menu. Given a query, three processes take place in a sequence of deduction, evaluation, and then explanation. In the deduction phase all deterministic goals and subgoals of the query are executed, and the deterministic part of the explanation tree is generated. The objective is to achieve efficiency in queries involving a set of solutions differing only in their nondeterministic parts of the query and of the explanation structure are evaluated.

ORBI's linguistic competence is achieved by means of a lexical and syntactic–semantic analysis, transforming a sentence into an optimally ordered directly evaluable list of Prolog goals (Figure 5). The lexical analysis replaces each word of the input sentence by its corresponding lexical category. The syntactic–semantic analysis is realized by means of a core grammar containing context-free and sensitive rules expressed in the definite clause grammar formalism (see Warren and Pereira 1982) with syntactic and semantic checks. The core grammar is independent of its application and is transportable to other domains. The core grammar is completed by a domain dependent module containing structures, vocabulary, and semantic controls pertaining to the domain of expertise.

Pereira et al. (1982) devoted much effort to incorporating elliptic and extraposed structures. Syntactic and semantic controls verify number and gender agreements, designations of complex entities, and compatibility between nouns and verbs and their complements, pointing out any faults. Ellipses occurring sentence to sentence are solved according to the structure of the previous sentence in the dialogue. For each grammatical sentence accepted it records morphological sequences that could be an ellipsis in the next sentence of the dialogue. If the next sentence is not elliptical the information is forgotten. If it is elliptical then the gaps of the sentence are filled with the different strings recorded in prespecified ways (Figure 5). The technique comprises transforming the elliptical sentence into a complete sentence and then analyzing it as a complete sentence.

ISSUES AND CHALLENGES

The fundamental challenge in SpQL is to provide a unified model for the retrieval and manipulation of geographic data. In meeting this challenge there are a number of issues to be addressed.

Underlying Data Models

Two closely coupled challenges will be development of an effective GIS–user dialogue interface that incorporates a language for spatial analysis. Historically the underlying data model of the particular system can have a fundamental impact on the SpQL. For example, QPE-based SpQLs have tended to be built upon a raster-based model while SQL-based SpQLs have tended to be built upon a vector-based model. Although there are transforms that allow systems to map into and from

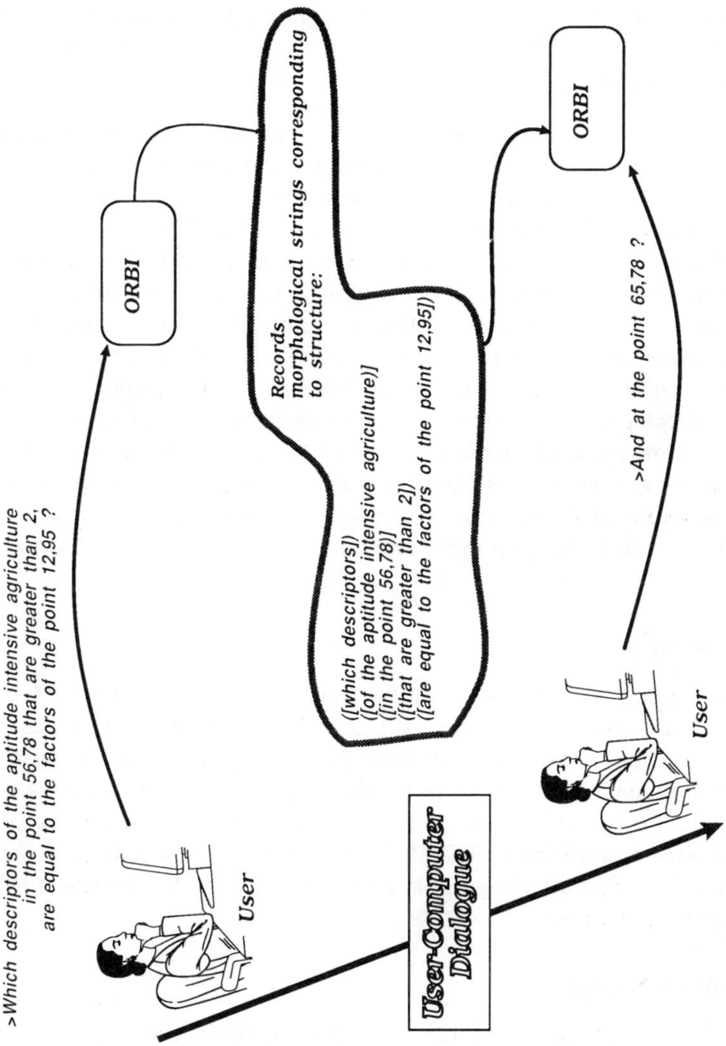

FIGURE 5. Example of a Natural Language Interface with some linguistic competence (Source: Pereira *et al.* 1982)

these very different models the process is not without its problems, especially in regards to issues of accuracy and precision. Current SpQLs are usually tightly coupled to the model upon which the system is based for that model determines how the SpQL is structured and the meaning of its vocabulary. One of the challenges of SpQL is therefore to provide a unified model of geographic information upon which a unified approach to the retrieval and manipulation of geographic data can be based.

Object-based Approach

It is clear that the basic idea of objects and their importance in SpQL has and will continue to exert substantial influence on the evolution of SpQLs. A common assumption is that users tend to organize their model of geographic information and processing around the idea of geographic objects. Thus there is growing interest and work in object-based and object-oriented approaches to SpQL. There are two architectures being used in this approach. One attempts to build an object-based SpQL as an interface layer between the user and the nonobject-oriented underlying geographic database system. On the surface this appears to be a more practical approach. The other approach includes an object-based SpQL as an integral part of an object-oriented GIS. The latter approach is a more unified model that is based on developments in object-oriented programming and database systems. Issues such as multiple inheritance (Mohan and Kashyap 1988) and management of persistent objects remain to be fully addressed.

Semantics

The syntax of SpQL is a simple matter compared with implementing a unified approach to the semantics of GIS. Simple problems can pose unexpected complexity. For example, specifying the meaning of a spatial relation such as near has shown that it is anything but a simple matter for SpQL to handle in a meaningful manner (Robinson et al. 1986*a*, Robinson and Wong 1987, Robinson 1990*b*). The challenge to SpQL is to be able to handle geographic semantics in something other than a rigid, arbitrary manner.

Self Knowledge

As illustrated by capabilities of the NLI of ORBI (Pereira et al. 1982) there is a need for an SpQL system to have knowledge of itself. A common, implicit assumption is that the GIS user has substantial knowledge of how to communicate with the GIS for a particular problem. The challenge here is to provide natural ways of relieving GIS users of cognitive and memory demands that impede their access to retrieval and manipulation capabilities of GIS. The use of expert systems, or applied AI, technology to resolve this issue has been suggested in a number of papers reviewing expert systems and GIS (Robinson and Frank 1987, Robinson et al. 1988, Robinson et al. 1989).

QL and Interoperability

Developments in SpQL and the increasing demand for geographic data exchange have begun to raise an interesting issue regarding the use of SpQL. Increasingly, the user of a GIS is another GIS (Figure 1). Standards for interchange have historically focused on formats, file structure, etc. But in the future there will be need for GISs to communicate and interact. Thus, one of the issues will be whether or not to base an inter-GIS QL upon that used by human users. Regardless of the choice of approach, the challenge will be to develop languages that will support GIS interoperability so that the full geographic information resources of GISs can be shared among a network of users and their GISs. This cluster of challenges is an outgrowth of the trend towards distributed workstation computing environments (Robinson 1990*c*).

Integration into Interface Dialogue

To improve access to GISs the SpQL will increasingly be integrated with a user interface that incorporates linguistic as well as nonlinguistic tools supporting the retrieval and manipulation of geographic information. There are numerous models of human–computer interface management (Hartson and Hix 1989). Which are suited for adaptation to the GIS environment will become an issue. The challenge will be to specify the requirements for human–GIS interface management and how it may, or may not, differ from the general human–computer interface management systems. In the near term a SpQL may be the language for representing and prototyping such interfaces. Portions of the human–GIS interface would moderate dialogue that would automate the construction of domain-specific aspects of a SpQL. This is an important issue in developing the future SpQL since experimental results have shown that user concepts of even simple spatial relations vary according to context (Robinson and Wong 1987).For a conventional query language, queries are specified only in terms of the defined vocabulary of the database.

PRACTITIONER'S SUMMARY

For the practising scientist and professional there are a number of accomplishments in SpQL that are worthy of attention. These accomplishments are primarily related to developments in Structured Query Language (SQL) extensions, object-based approaches, and spin-offs of natural language developments.

SQL

A number of GIS products support some form of extension to SQL which permits query operations on geographic data. This accomplishment is significant in several respects. It provides a well-understood link to other, nongeographic, database systems. Although there is no accepted standard set of extensions to SQL for GIS query operations, the extensions are contributing to development of a commonly understood geographic QL as a subset of SQL.

Object-based

The object-based approach to user–system interface and retrieval of geographic information clearly increases the usability of SpQL. However, several fundamental issues are yet to be resolved when extending the object-based approach to the full range of operations required by a GIS. In some cases performance problems will diminish with advances in memory and processor technology.

Natural Language

Simple natural language interfaces to research-oriented systems have been developed. However, the obstacles to fielding a full NL-based SpQL for GIS remain formidable. As long as the primary devices for interaction remain keyboard/mouse oriented the utility of an NL approach will be questionable. With advances in speech recognition/synthesis there will be a need for NL-based SpQL for GIS. Spinoffs from NL/SpQL research will continue to affect other areas of SpQL, most notably development of a GIS extension to SQL.

Integration

In the near-term those who are trying to use, deliver, support, and apply GIS will be primarily concerned with the use of existing tools and languages. Most of these efforts are occurring within an organizational context so that the influence of database QLs will increase as GIS becomes integrated with the remainder of the organization. The pressures of integration with organizational information resources and limits of user willingness to deal with the increased complexity suggest two broad elements of a practitioners research agenda:

Geographic Extensions to Database Query Languages. A consensus is evolving for development of standard database query languages such as SQL by organizations in which GIS is a component of a larger set of corporate information resources. This trend has already affected GIS

(Robinson 1990*b*). Although many approaches already use SQL for nonspatial, or attribute, data management there will be increasing pressure for incorporation of GIS-related extensions to languages such as SQL.

Interface Development SpQLs will become increasingly integrated with GIS–user interfaces. GIS interfaces will need to be designed to reduce the cognitive burden on the GIS user and be integrated into a general database interface. The implication here is that current tools will be used to increase the GIS competence of SpQL. Integration of SpQL with other interface management tools means that general trends in query languages, hardware, and software will have a fundamental impact on how these interfaces are constructed.

A common assumption of GIS–user interface being the sole focus of SpQL will become less dominant as the area of spatial data interchange increases in importance. SpQL will take on increased significance as a mechanism for GIS–GIS interaction.

PROJECTIVE SUMMARY

A research agenda for development of the future generation SpQL should include:

Systematic research into the human factors of SpQL. This body of research should be the cornerstone upon which future SpQLs are based.

For progress in this area a symbiotic research relationship is needed. For example, while management-level concerns increasingly focus on human factors, especially interaction technologies, technical professionals have shown little predilection to go beyond introspection and their own experience. Casual introspection hardly seems an adequate basis to develop costly and widely used computer and information systems (Shneiderman 1978). Yet, most current SpQL development is based on occasional introspection. One of the few cases of using an experimental technique in SpQL research is presented by Robinson and Wong (1987) in their work on specifying spatial relation terms.

Development of a unified model of geographic information. This would provide a commonly understood model upon which to base a SpQL that supports retrieval and manipulation.

Object-oriented database systems (Drabowski et al. 1990) hold the potential to provide a unified model that can accommodate the query, retrieval, and manipulation of geographic data in a manner meaningful to GIS users.

Increase the NL competence of SpQL. This will become increasingly important as innovative technologies are integrated with information systems' interfaces.

Although natural language may not be appropriate for all users, there are significant research issues that could increase the utility of NL as a SpQL. As technical capabilities for speech recognition as an input channel develop, voice input/output interface will have a profound impact (Straub and Wetherbe 1989) on how we interact with GIS. To support this interface, NLI will be required that have some increased measure of linguistic competence, especially in regards to the linguistics of spatial relations. NL for query in GIS requires considerable GIS competence in order to interpret a user's query. Intelligent, expert systems/applied AI interfaces that permit users to express their needs in a form more natural to them will be critical in meeting the overall objective of providing GISs that have a high level of geographical competence while not burdening the user with undue cognitive load (Robinson and Frank 1987, Lundberg and Robinson 1988, Robinson et al. 1989).

Increase ability of SpQL to handle semantic variations across domains. The handling of domain-specific semantics will be needed in future SpQL.

A common problem in establishing geographic extensions to SQL, and other languages, has been the difficulty in dealing with variations in the meaning of terms across and within domains. This problem was illustrated by the substantial inter-user variation found in a study of user specification of a simple spatial relation (Robinson and Wong 1987). The difficulty of handling these domain-specific problems has been mentioned by SpQL developers (eg Herring et al. 1988, Orenstein and Manola 1988) and handled through tools for making their systems extensible. In the future, tools for formalizing domain-specific extensions will exploit system and domain knowledge to ease the burden of extension programming on those organizations with substantial investments in geographic information technology.

Incorporation of artificial intelligence in SpQL operations.

Future systems will be an active rather than passive partner in interpreting user requests (Straub and Wetherbe 1989). A primitive example of such an active system is the Spatial Relations Acquisition Station (SRAS) described by Robinson (1990*b*) and Robinson and Wong (1987). SRAS develops the specification of a spatial relation for use in a fuzzy spatial query through an active interaction with the user.

A study of information technologies (IT) for the 1990s revealed that organizations consider *human interface technologies* (HIT) as a critical group of IT. HIT is considered to be a critical factor in the future impact of IT within organizations as it is the technology bundle that enhances or diminishes access to an organization's information resources (Straub and Wetherbe 1989). SpQL should be considered a specialized component of the HIT group. As such, developments in SpQL will continue to exert considerable influence over the degree of impact that GIS has in society.

REFERENCES AND BIBLIOGRAPHY

Abel D J, Smith J L (1986) A relational GIS database accommodating independent partitionings of the region. In: *Proceedings of the Second International Symposium on Spatial Data Handling*: 213–24

American National Standards Institute (1986) *American National Standard for Information Systems — Database Language — SOL.* ANSI X3. 135–1986

Antonacci F, Bartolo L, Orco P, Spadavecchia N (1979) AQL: A relational database management system and its geographical applications. In: Blaser A (ed.) *Data Base Techniques for Pictorial Applications.* Springer-Verlag: 569–99

Baecker R M, William A, Buxton S (1987) *Readings in Human–Computer Interaction: A Multidisciplinary Approach.* Los Altos, California, Morgan Kaufmann

Barrera R, Buchmann A (1981) Schema definition and query language for a geographical database system. In: *Proceedings IEEE Conference on Computer Architecture for Pattern Analysis and Image Database Management*: 250–6

Chang N S, Fu K S (1980) Query-by-pictorial-example. *IEEE Transactions on Software Engineering* **SE – 6** (6): 519–24

Chang N S, Fu K S (1981) Picture query languages for pictorial data-base systems *Computer* **14** (11): 23–33

Chang S-K, Kunii T L (1981) Pictorial data-base systems. *Computer* **14** (11): 13–17

Chang S-K, Reuss J, McCormick B H (1978) Design considerations of a pictorial database system. *Policy Analysis and Information Systems* **1** (2): 49–70

Chang S-K, Lin B S, Walser R (1979) A generalized zooming technique for pictorial data base systems. In: *AFIPS Conference Proceedings* **48**: 147–56

Charlwood G, Moon G, Tulip J (1987) Developing a DBMS for geographic information: A review. In: *Proceedings Eighth International Symposium on Computer-Assisted Cartography.* Baltimore, Maryland: 302–15

Chock M (1981) Manipulating data structures in pictorial information systems *Computer* **14** (11): 13–17

Codd E F (1970) A relational model of data for large shared data banks. *Communications of Association of Computing Machinery* **13** (2): 377–97

Date C J (1983) *Database: A Primer.* Reading, Massachusetts, Addison-Wesley

Drabowski C E, Fong E N, Yang D (1990) *Object Database Management Systems: Concepts and Features,* National Institute of Standards and Technology Special Publication 500–179. Washington, DC, US Government Printing Office

Foley J D, Wallace V L (1974) The art of natural graphic man–machine conversation. *Proceedings IEEE* **63** (4): 462–71

Frank A U (1982) MAPQUERY: Data base query language for retrieval of geometric data and their graphical representation. *Computer Graphics* **16** (3): 199–207

Goh P-C (1989) A graphic query language for cartographic and land information systems. *International Journal of Geographic Information Systems* **3** (3): 245–55

Herot C F (1980) Spatial management of data. *ACM Transactions on Database Systems* **5** (4): 493–513

Hartson H R, Hix D (1989) Human–computer interface development: Concepts and systems. *ACM Computing Surveys* **21** (1): 5–92

Herring J R (1987) TIGRIS: Topologically integrated geographic information system. In: *Proceedings Eighth International Symposium on Computer-Assisted Cartography,* Baltimore, Maryland: 282–91

Herring J R, Larsen R C, Shivakumar J (1988) Extensions to the SQL language to support spatial analysis in a topological data base. In: *Proceedings GIS/LIS'88* **2**: 741–50

Ingram K J, Phillips W W (1987) Geographic information processing using an SQL-based query language. In: *Proceedings Eighth International Symposium on Computer-Assisted Cartography*, Baltimore, Maryland: 326–35

Jarke M, Vassiliou Y (1985) A framework for choosing a database query language. In: Mylopoulos J and Brodie M L (ed.), *Readings in Artificial Intelligence & Databases*. San Mateo, California, Morgan Kaufmann: 363–76

Joseph T, Cardenas A F (1988) PICQUERY: A high level query language for pictorial database management. *IEEE Transactions on Software Engineering* **14** (5): 630–8

Kasturi R, Fernandez R, Amlani M L, Feng W (1989) Map data processing in geographic information systems. *Computer* **22** (12): 10–21

Keating T, Phillips W, Ingram K (1987) An integrated topologic database design for geographic information systems. *Photogrammetric Engineering and Remote Sensing* **53** (10): 1399–1402

Larson J A (1987) User interfaces to database management systems. In: Larson J A (ed.) *Database Management Tutorial*. Washington, DC. IEEE Computer Society Press: 83–101

Lien W Y, Harris S K (1980) Structured implementation of an image query language. *Pictorial Information Systems* **3**: 416–30

Lin B S, Chang S-K (1980) GRAIN — A pictorial database interface. In: *Proceedings of IEEE Workshop on Picture Data Description and Management*: 83–8

Lundberg G, Robinson V B (1988) Computers that know: A tutorial. *Computers, Environment, and Urban Systems* **12** (1): 49–72

Malone T (1982) Heuristics for designing enjoyable user interfaces: Lessons learned from computer games. In: *Proceedings of the Conference on Human Factors in Computer Systems*, Gaithersburg, Maryland: 63–8

Manola F, Orenstein J, Dayal U (1987) Geographic information processing in the PROBE database system. In: *Proceedings Eighth International Symposium on Computer-Assisted Cartography*, Baltimore, Maryland: 316–25

Mantey P E, Carlson E D (1979) Integrated geographic data bases. In: Blaser A (ed.), *Data Base Techniques for Pictorial Applications*. Amsterdam, Springer-Verlag: 173–98

McCann C A, Taylor M M, Touri M I (1988) ISIS: The interactive spatial information system. *International Journal of Man–Machine Studies* **28**: 101–38

Menon S, Smith T R (1989) A declarative spatial query processor for geographic information systems. *Photogrammetric Engineering and Remote Sensing* **55** (11): 1593–1600

Mohan L, Kashyap R L (1988) An object-oriented knowledge representation for spatial information. *IEEE Transactions on Software Engineering* **14** (5): 675–80

Nagy G, Wagle S(1979) Geographic data processing. *ACM Computing Surveys* **11** (2): 139–81

Oberg R E (1988) The visual cartographic enquiry system for geodetic control data. In: *GIS/LIS'88 Proceedings: Accessing the World*. San Antonio, Texas: 484–9

Orenstein J A, Manola F A (1988) PROBE spatial data modeling and query processing in an image database application. *IEEE Transactions on Software Engineering* **14** (5): 611–29

Pereira L M P, Sabatier P, de Oliveira E (1982) *ORBI — an expert system for environmental resource evaluation through natural language.* Report FCT/DI – 3/82, Departmento de Informatica, Universidade Nove de Lisboa (presented at First International Conference on Logic Programming, Marseilles, France)

Pizano A, Klinger A, Cardenas A (1989) Specification of spatial integrity constraints in pictorial databases. *Computer* **22** (12): 59–71

Reisner P (1981) Human factors studies of database query languages: A survey and assessment. *ACM Computing Surveys* **13** (1): 13–31

Retz-Schmidt G (1988) Various views on spatial prepositions. *AI Magazine* **9** (2): 95–105

Robinson V B (1980) *Urban Data Management Software (U.D.M.S.) Package — User's Manual (Draft).* Nairobi, Kenya, United Nations Centre for Human Settlements (HABITAT).

Robinson V B (1990*a*) *Structured query language (SQL) and geographic information systems (GIS): A briefing review,* Research Agreement Report No. 1 (Draft), GIS Standards Laboratory, National Institute for Standards and Technology, United States Department of Commerce

Robinson V B (1990*b*) Interactive machine acquisition of a fuzzy spatial relation. *Computers and Geosciences* (in press)

Robinson V B (1990*c*) Intelligent geographic information systems: AI, DB, and GIS: Technologies working together (plenary address abstract). In: *Proceedings of GIS for the* 1990's *National Conferences,* Ottawa, Ontario (in press)

Robinson V B, Coiner J C (1986) Characteristics and diffusion of a microcomputer geoprocessing software package: The urban data management software package. *Computers, Environment, and Urban Systems,* **10** (3/4): 165–73

Robinson V B, Frank A U (1985) About different kinds of uncertainty in geographic information systems. In: *Proceedings Seventh International Symposium on Computer-Assisted Cartography,* Washington, DC: 440–50

Robinson V B, Frank A U (1987) On expert systems for geographic information systems. *Photogrammetric Engineering and Remote Sensing* **53** (10): 1435–41

Robinson V B, Wong R N (1987) Acquiring approximate representations of some spatial relations. In: *Proceedings Eighth International Symposium on Computer-Assisted Cartography,* Baltimore, Maryland: 604–22.

Robinson V B, Blaze M A, Thongs D (1985) Natural language in geographic data processing. In: *Proceedings International Conference of the Remote Sensing Society and Center for Earth Resources Management,* London: 67–73

Robinson V B, Blaze M A, Thongs D (1986*a*) Representation and acquisition of a natural language relation for spatial information retrieval. In: *Proceedings Second International Symposium on Spatial Data Handling,* Seattle, Washington: 472–87

Robinson V B, Blaze M A, Thongs D (1986*b*) Man–machine interaction for acquisition of spatial relations as natural language concepts. In: Optiz B K (ed.) *Geographic Information Systems in Government.* Hampton, Virginia, A. Deepak Publishing: 433–54

Robinson V B, Frank A U, Blaze M A (1988) Expert systems and geographic information systems: Review and prospects. In: Pham D T (ed.) *Expert Systems in Engineering*. AI in Industry Series. Kempston, Bedford, UK, IFS (Publications) Ltd: 203–13

Robinson V B, Frank A U, Karimi H A (1989) Expert systems for geographic information systems in natural resource management. In: Ripple W J (ed.) *Fundamental Geographic Information System Concepts: A Compendium*. Bethesda, Maryland, American Society of Photogrammetry and Remote Sensing: 155–66

Roussopoulos N, Faloutsos C, Sellis T (1988) An efficient pictorial database system for PSQL. *IEEE Transactions on Software Engineering* **14** (5): 639–50

Samet H (1990) QUILT: A geographic information system based on quad-trees. *International Journal of Geographical Information Systems* **4** (2): 103–32

Samet J (ed.) (1981) Query languages — a unified approach. Report of the British Computer Society Query Languages Group. Cambridge, UK, Heyden University Press.

Shneiderman B (1978) Improving the human factors aspect of database interactions. *ACM Transactions on Database Systems* **3** (4): 417–39

Small D W, Weldon L J (1983) An experimental comparison of natural and structured languages. *Human Factors* **25** (3): 253–63

Straub D W, Wetherbe J C (1989) Information technologies for the 1990's: An organizational impact perspective. *Communications of the ACM* **32** (11): 1328–39

Tanaka M, Ichikawa T (1988) A visual user interface for map information retrieval based on semantic significance. *IEEE Transactions on Software Engineering* **14** (5): 666–70

Tanaka S, Ikeda H (1989) An integrated data model for statistical and geographical data management. In: Kyung W L and Wu C (ed.) *Proceedings of World Conference on Information Processing/Communication*, June 14–16: 444–51

Tuori M, Moon G (1984) A topographic map conceptual data model. In: *Proceedings International Symposium on Spatial Data Handling*, Zurich, **1**: 28–37

Vassiliou Y, Jarke M (1984) Query languages — a taxonomy. In: Vassiliou Y (ed.) *Human Factors and Interactive Computer Systems*. Norwood, New Jersey, Ablex: 47–81

Warren, D H D, Pereira, C N (1982) An efficient easily adaptable system for interpreting natural language queries. *Am. J. Computational Linguistics* **8** (3/4): 110–23

Waugh T C, Healey R H (1986) The GEOVIEW design: A relational database approach to geographical data handling. In: *Proceedings, Second International Symposium on Spatial Data Handling*, Seattle, Washington: 193–212

Welty C, Stemple D W (1981) Human-factors comparison of a procedural and a non-procedural query language. *ACM Transactions on Database Systems* **6** (4): 626–49

Westwood K A, Brinkman J P (1988) Toward the successful integration of relational and quadtree structures in a geographic information system In: Salling M (ed.) *Proceedings Urban and Regional Information Systems Association*: 181–89

Wu J-K, Chen T, Yang L (1989) QPF a versatile query language for a knowledge-based geographical information system. *International Journal of Geographic Information Systems* **3** (1): 51–9

Zloof M M (1975) Query-by-example. In: *AFIPS Conference Proceedings* **44**: 431–8

Zobrist A L, Nagy G (1981) Pictorial information processing of Landsat data for geographic analysis. *Computer* **14** (11): 34–41

Chapter 6

ISSUES OF QUALITY AND UNCERTAINTY

Michael F. Goodchild
National Center for Geographic Information and Analysis, University of California, Santa Barbara, CA 93106, USA

INTRODUCTION

Digital processing of cartographic data brings immense benefits in the form of rapid, precise and sophisticated analysis, but at the same time it reveals weaknesses which may not otherwise be apparent. Computers are very precise machines, and errors and uncertainties in data can lead to serious problems, not only in the form of inaccurate results but in the consequences of decisions made on the basis of poor data, and increasingly in legal actions brought by affected parties. In this chapter we first review the dimensions of the accuracy problem, and discuss the premises and assumptions on which the remainder of the chapter is based. The second section reviews the models which are available for analysing and understanding errors in spatial data. The chapter concludes with a summary of the current state of the art and an agenda for future research and development.

The terms accuracy, precision, resolution and scale are used almost interchangeably in reference to spatial data, but we will need to establish more exact definitions for the purposes of this discussion. *Accuracy* refers to the relationship between a measurement and the reality which it purports to represent. *Precision* refers to the degree of detail in the reporting of a measurement or in the manipulation of a measurement in arithmetic calculations. Finally, the *resolution* of a data set defines the smallest object or feature which is included or is discernable in the data. Scale and resolution are intimately related because there is a lower limit to the size of an object which can be usefully shown on a paper map. This limit is often assumed to be 0·5mm as a rule of thumb, so the effective

TABLE 1 Scale and Resolution for Some Common Map Scales

Scale	*Effective resolution*
1:1250	62·5 cm
1:10000	5 m
1:24000	12 m
1:50000	25 m
1:100000	50 m
1:250000	125 m
1:500000	250 m
1:1000000	500 m
1:10000000	5 km

resolution of a 1: 1000 map is about 1000 times 0·5mm, or 50cm, although the standards of most mapping agencies are substantially better. The effective resolutions of some common map scales, assuming 0·5mm resolution, are shown in Table 1.

Because digital spatial data sets are not maps, and have no similar physical existence, they do not have an obvious scale, and it is common to think of such data as essentially scale-free. But this is to oversimplify. If the data were obtained by digitizing, their resolution is that of the digitized map. If they were obtained from imagery, their resolution is that of the pixel size of the imagery. In spite of appearances to the contrary, then, no spatial data are ever resolution free. The term 'scale' is often used loosely as a surrogate for database resolution.

DIMENSIONS OF THE PROBLEM

Precision

One of the widely accepted design principles of spatial data handling systems is that they should be able to process data without significant distortion. It is important when calculations are carried out on co-ordinates, for example, that the results should not be affected by lack of precision in the machine. We have seen that a figure of 0·5mm can be assumed to be typical for the resolution of an input map document. 0·5mm is also roughly the accuracy with which an average digitizer operator can position the crosshairs of a cursor over a point or line feature. Most digitizing tables have a precision of 0·25mm or better, thus ensuring that the table itself does not introduce additional distortion as points are captured.

If we take the size of the source map sheet on the digitizing table to be

100cm by 100cm, then a precision of 0·5mm represents an uncertainty of 0·05% relative to the size of the sheet, or an uncertainty in the fourth significant digit. So in order not to introduce distortion, internal storage and processing of coordinates must be correct to at least four digits. However, seven or eight decimal digits are commonplace in arithmetic calculations in spatial data handling systems, and many GISs operate at much higher precision, in effect allowing GIS users to ignore the possibility of arithmetic distortion and to assume that internal processing is carried out with infinitely high precision.

Despite this, arithmetic precision, or lack of it, is sometimes a problem in spatial data processing. One common situation occurs when point locations are specified in a global coordinate system. For example, the UTM system assigns coordinates with respect to an origin on the equator, so locations in the mid-Northern latitudes will have one of their coordinates in the millions of metres. Processing of data taken from large scale maps, perhaps 1: 1 000, will therefore require precision in the sixth or seventh significant digit. This can be particularly problematic in performing operations such as calculation of the intersection point of two lines, or the centroid of a parcel of land, where differences in the products of six or seven digit numbers become important.

Another artifact of high precision is well known in connection with polygon overlay operations. It is common to find that two coverages of different themes for the same area share certain lines. For example the shoreline of a lake will appear both on a map of soils and a map of vegetation cover. When soils and vegetation cover are overlaid the two versions of the shoreline should match perfectly. In practice this is never the case. Even if the two lines matched perfectly on the input documents, errors in registration and differences in the ways in which the digitizer operators captured the lines will ensure different digital representations in the database. Moreover since the overlay operation is carried out to high precision, the differences are treated as real and become small sliver polygons. Table 2 (from Goodchild 1979), shows the results of overlaying five layers from the Canada Geographic Information System, and the enormous numbers of very small polygons produced.

Several approaches have been taken to deal with the spurious polygon problem in polygon overlay, which has the annoying property that greater accuracy in digitizing merely leads to a greater number of slivers. Some systems attempt to distinguish between spurious and real polygons after overlay using simple rules, and delete those which are determined to be spurious (Figure 1). A suitable set of rules to define spurious polygons might be:

- small in area;
- long and thin in shape (high ratio of perimeter to square root of area);

TABLE 2 Polygons by Area for Five CGIS Layers and Overlays

Acres	*1*	*2*	*3*	*4*	*5*	*1+2+5*	*1+2+3+5*	*1+2+3+4+5*
0–1	0	0	0	1	2	2640	27 566	77 346
1–5	0	165	182	131	31	2195	7521	7330
5–10	5	498	515	408	10	1421	2108	2201
10–25	1	784	775	688	38	1590	2106	2129
25–50	4	353	373	382	61	801	853	827
50–100	9	238	249	232	64	462	462	413
100–200	12	155	152	158	72	248	208	197
200–500	21	71	83	89	92	133	105	99
500–1000	9	32	31	33	56	39	34	34
1000–5000	19	25	27	21	50	27	24	22
> 5000	8	6	7	6	11	2	1	1
Totals	88	2327	2394	2149	487	9558	39 188	90 599

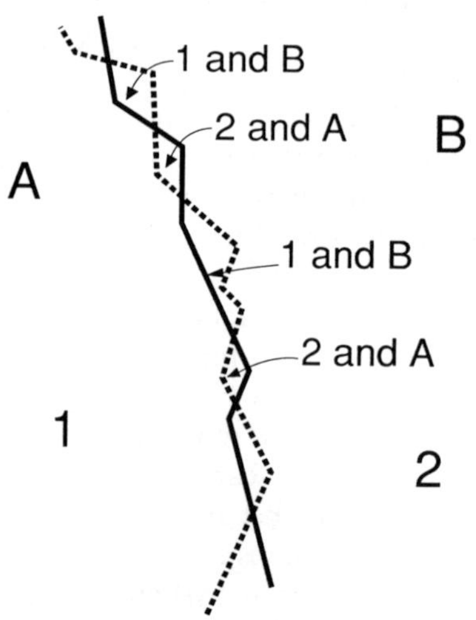

FIGURE 1 Generation of sliver polygons in overlay

- composed of two arcs only;
- both nodes have exactly four incident arcs.

In addition, the attributes of spurious polygons are characteristic. Suppose our soil map line marks the boundary between soils A and B, and our vegetation map shows the boundary between classes 1 and 2. Ideally, when overlaid the result should be a line with soil A and vegetation 1 on one side, and soil B and vegetation 2 on the other. Sliver polygons will be either soil B and vegetation 1, or soil A and vegetation 2. Moreover the

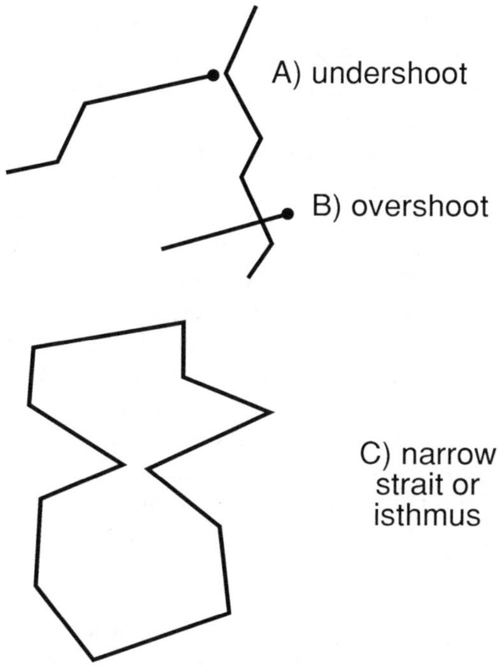

FIGURE 2 Tolerance problems in digitizing

attributes will alternate from one sliver to the next, as the soil and vegetation boundaries cross and recross (Figure 1). So the detection of one spurious polygon provides additional evidence that nearby polygons are also spurious.

Another approach to dealing with the sliver problem is in effect to reduce the precision of the overlay operation, by requiring the user to establish a tolerance distance. Any lines lying within the tolerance distance of each other are assumed to be the same. A similar approach is often taken in digitizing, where two lines are assumed to join or 'snap' if the distance between them is less than some user-established tolerance distance (see Figure 2A).

Because a tolerance distance is in effect a reduced level of spatial resolution, all objects or features smaller than the tolerance become ambiguous or indistinguishable. Figure 2C shows an example of a complex polygon with a narrow strait or isthmus. If the width of the strait is less than the tolerance distance, it is impossible to determine from the geometry of the feature alone whether it is indeed a single polygon with a narrow strait, or two polygons. In some cases it may be possible to resolve the ambiguity if the user has already identified labels or attributes for the polygon(s); if two labels have been input, it should be clear that the feature consists of two polygons. In such instances the *topology* (the existence of two polygons) is allowed to resolve ambiguity in the *geometry*.

The potential conflict between topology and geometry is a consistent theme in spatial data handling. Franklin (1984) and Franklin and Wu (1987) have written about specific instances of conflict, and about ways of resolving them. For example, a point may be given an attribute indicating that it lies inside a given polygon, which might be a county, but geometrically, because of digitizing error or other problems, the point may lie outside the polygon (Blakemore 1984). The conflict may be resolved by moving the point inside the polygon, or by moving the polygon boundary to include the point, or by changing the point's attributes. The first two are examples of allowing topology to correct geometry, and the third of allowing geometry to correct topology. In all cases, and particularly in the second, the implications of a change on other objects and relationships must be considered; if the polygon's boundary is moved, do any other points now change properties? Unfortunately the propagation of changes of this type can have serious effects on the overall integrity of the database.

In summary, the average user will approach a GIS or spatial data handling system with the assumption that all internal operations are carried out with infinite precision. High precision leads to unwanted artifacts in the form of spurious polygons. Moreover in reducing effective precision in such operations as digitizing and polygon closure it is common for unwanted effects to arise in the form of conflicts between the topological and geometrical properties of spatial data. The ways in which system designers choose to resolve these conflicts are important to the effective and data-sensitive application of GIS technology.

Accuracy

Thus far we have used examples of the distortions and errors introduced into spatial data by digitizing and processing. However if we are to take a comprehensive view of spatial data accuracy it is important to remember that accuracy is defined by the relationship between the measurement and the reality which it purports to represent. In most cases reality is not the source document but the ground truth which the source document models. The map or image is itself a distorted and abstracted view of the real world, interposed as a source document between the real world and the digital database.

We will use two terms to distinguish between errors in the source document and in the digitizing and processing steps. *Source* errors are those which exist in the source document, and define its accuracy with respect to ground truth. *Processing* errors are those introduced between the source document and the GIS product, by digitizing and processing.

We will find that, in general, processing errors are relatively easier to measure, and smaller than source errors.

All spatial data without exception are of limited accuracy, and yet it is uncommon to find statements of data quality attached to such data, whether in the form of source maps, images or digital databases. The accuracy of much locational or attribute data is limited by problems of measurement, as for example in the determination of latitude and longitude, or elevation above mean sea level. In other cases accuracy is limited by problems of definition, as when a soil type is defined using terms such as 'generally', 'mostly', 'typically' and thus lacks precise, objective criteria. Locations defined by positions on map projections are limited by the accuracy of the datum, and may change if the datum itself is changed.

More subtle inaccuracies result from the way in which a source document models or abstracts reality. For example an extended object such as a city may be shown at some scales as a point, or a census reporting zone may be located by its centroid or some other representative point (Bracken and Martin 1989). The transition between two soil types, which occurs in reality over some extended zone, may be represented on the source document and in the database by a simple line or sharp discontinuity. The attributes assigned to a polygon may not in fact apply homogeneously to all parts of the polygon, but may be more valid in the middle and less valid toward the edges (Mark and Csillag 1989). All of these errors result from representing spatial variation using objects; the objects are either of the wrong type for accurate representation, in the case of the city represented by a point, or used to model what is in fact continuous variation. Models such as these allow complex spatial variation to be expressed in the form of a comparatively small number of simple objects, but at a corresponding cost in loss of accuracy. Unfortunately the process of modelling has usually occurred in the definition of the source document, long before digitizing or processing in a GIS, and rarely is any useful information on levels of accuracy available.

Separability

Most digital spatial data models distinguish clearly between locational and attribute information. From an accuracy perspective, errors in location and attributes are often determined by quite different processes, and in these cases we term the two types of error *separable*. For example, a reporting zone such as a census tract may be formed from segments which follow street centre lines, rivers, railroads and political boundaries. The processes leading to error in the digitizing of a river-based boundary are quite different to those typical of street centre lines. Errors in imagery are similarly separable, attribute errors being determined by problems of

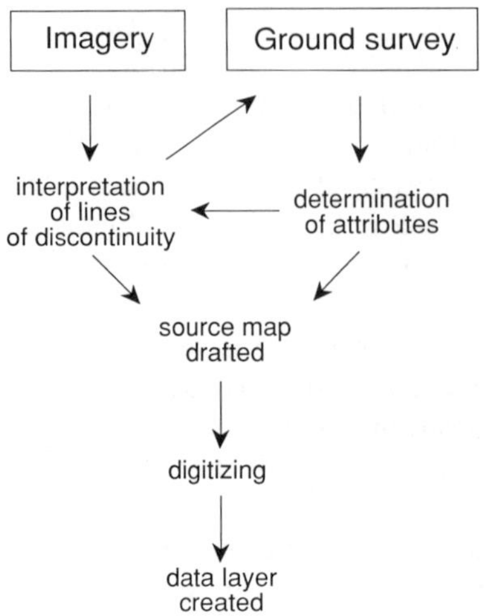

FIGURE 3 Stages in the creation of a forest cover layer

spectral response and classification, while locational errors are determined by problems of instrument registration. The same is true to a lesser extent for topographic data, since errors in determining elevation at a point are probably only weakly dependent on errors in locating the point horizontally.

However in maps of vegetation cover, soils or geology, where polygon objects are commonly used to model continuous variation, the two types of error are not separable, as object boundaries are derived from the same information as the attributes they separate (Goodchild 1989). The transition zone between types A and B may be much less well defined than the transition between types A and C. The process by which such maps are created gives useful clues to the likely forms of error which they contain. A forest cover map, for example, will be derived from a combination of aerial imagery and ground truth. Polygons are first outlined on the imagery at obvious discontinuities. Ground surveys are then conducted to determine the attributes of each zone or polygon. The process may be iterative, in the sense that the ground survey may lead to relocation of the interpreted boundaries, or the merger or splitting of polygons. Figure 3 summarizes these processes of source error generation, and the subsequent steps which lead to processing errors.

The forest example illustrates the relationship between different error processes particularly well. In forest management the homogeneous zones identified in the mapping process are termed *stands*, and become the basic unit of management. If a stand is cut and reseeded or allowed to regenerate, the processes determining its attributes may be no longer related to the processes determining its boundaries, and thus attribute and locational errors will become separable. Similarly, even in the initial map obtained from imagery and ground survey there will be boundary segments which follow roads, rivers or shorelines and again are subject to error processes which are independent of those affecting attributes.

The Ideal

The arguments in the preceding sections lead to a clear idea of how an ideal GIS might be structured. First, each object in the database would carry information describing its accuracy. Depending on the type of data and source of error, accuracy information might be attached to each primitive object, or to entire classes of objects. Accuracy might be coded explicitly in the form of additional attributes, or implicitly through the precision of numerical information.

Every operation or process within the GIS would track error, ascribing measures of accuracy to every new attribute or object created by the operation. Uncertainty in the position of a point, for example, would be used to determine the corresponding level of uncertainty in the distance calculated between two points, or in the boundary of a circle of specified radius drawn around the point.

Finally, accuracy would be a feature of every product generated by the GIS. Again it might be expressed either explicitly or implicitly in connection with every numerical or tabular result, and means would be devised to display uncertainty in the locations and attributes of objects shown in map or image form.

Of course the current state of GIS technology falls short of this ideal in all three areas. Currently we lack comprehensive methods of describing error, modelling its effects as it propagates through GIS operations, and reporting it in connection with the results of GIS analysis. The remaining sections of this chapter describe the current state of knowledge and such techniques as currently exist for achieving a partial resolution of the error problem.

MODELS OF ERROR

Background

With the basic introduction to accuracy and error in the previous section, we can now consider the specific issue of quality in cartographic data. This is not a simple and straightforward extension; as we will see, spatial data requires a somewhat different and more elaborate approach.

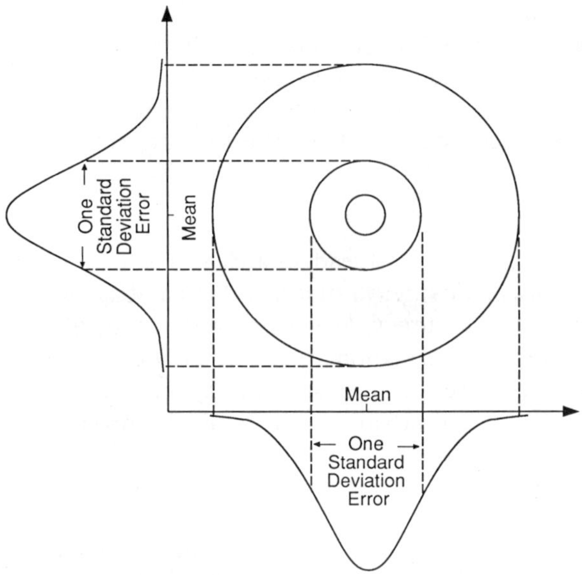

FIGURE 4 The circular normal model of errors in point locations

The objective is to find an appropriate model of the errors which occur in cartographic data. A model is taken here to mean a statistical process whose outcome emulates the pattern of errors observed in real digital spatial data. The first subsection considers models of error in the positioning of simple points; the second section looks at the ways this can be extended to more complex line or area features.

Points

The closest analogy between the classical theory of measurement and the problem of error in cartographic data concerns the location of a single point. Suppose, for example, that we wished accurately to determine the location, in coordinates, of a single street intersection. It is possible to regard the coordinates as two separate problems in measurement, subject to errors from multiple sources. If the coordinates were in fact determined by the use of a transparent roamer on a topographic sheet, then this is a fairly accurate model of reality. Our conclusion might be that both coordinates had been determined to an accuracy of 100m, or 2mm on a 1: 50 000 sheet. This is illustrated in Figure 4.

If the errors in both coordinates are represented by classic bell curves, then we can represent accuracy in the form of a set of ellipses centred on the point. If the accuracies are the same in each coordinate and if the errors are independent, then the ellipses become circles, again centred on the point. This gives us the circular normal model of positional error, as the two-dimensional extension of the classic error model. The error in position of a point is seen in terms of a bell-shaped surface centred over the point (Figure 4). The probability that any point is the true location is measured by the height of the surface of the bell over the point, and is clearly greatest at the measured location, decreasing symmetrically in all directions. Using the model we can compute the probability that the true location lies within any given distance of the measured location, and express the average distortion in the form of a standard deviation.

The *Circular Standard Error* is equal to the standard errors in the two coordinates, and can be portrayed by drawing a circle which passes through points representing one standard error to the left and right of the point in the x direction, and similarly above and below the point in the y direction (Figure 4). When the standard errors in x and y are not equal, the Circular Standard Error is approximated as the mean of the two. This approximation is valid only for small differences in the two standard errors, but it is difficult to imagine circumstances in which this would not be true. The probability that a point's true location lies somewhere within the circle of radius equal to the Circular Standard Error is 39·35%. The *Circular Map Accuracy Standard* (CMAS) requires that no more than 10% of the errors on a map will exceed a given value. For example, with a CMAS of 1mm, in order to meet the standard no more than 10% of points should be more than 1mm from their true positions. Under the circular normal model a radius of 2·146 times the circular standard error will contain 90% of the distribution, with 10% lying outside. The relationship between the CMAS and the circular standard error is therefore a ratio of 2·146.

These indices have been incorporated into many of the widely followed standards for map accuracy. For example, the NATO standards rate maps of scales from 1:25 000 to 1:5 000 000 as A, B, C or D as follows:

A: CMAS = 0·5mm (eg 12·5m at 1:25 000)
B: CMAS = 1·0mm (eg 25m at 1:25 000)
C: CMAS determined but greater than 1·0mm
D: CMAS not determined

Many other mapping programs use CMAS as the basis for standards of horizontal or planimetric accuracy, and use similar values.

Lines and Areas: The Perkal Band

In vector databases lines are represented as sequences of digitized points connected by straight segments. One solution to the problem of line error would therefore be to model each point's accuracy, and to assume that the errors in the line derived entirely from errors in the points. This is illustrated in Figure 5. Unfortunately this would be inadequate for several reasons. First, in digitizing a line a digitizer operator tends to choose points to be captured fairly carefully, selecting those which capture the form of the line with the greatest economy. It would therefore be incorrect to regard the points as randomly sampled from the line, or to regard the errors present in each point's location as somehow typical of the errors which exist between the true line and the digitized representation of it.

Secondly, the errors between the true and digitized line are not independent, but instead tend to be highly correlated (Amrhein and Griffith 1987, Keefer et al. 1988). If the true line is to the east of the digitized line at some location along the line, then it is highly likely that its deviation immediately on either side of this location is also to the east by similar amounts. Much of the error in digitized lines results from misregistration, which creates a uniform shift in the location of every point on the map. The relationship between true and digitized lines cannot therefore be modelled as a series of independent errors in point positions.

One commonly discussed method of dealing with this problem is through the concept of an error band. Figure 6 shows the observed line surrounded by a band of width epsilon, known as the Perkal epsilon band (Perkal 1956, 1966, Chrisman 1982, Blakemore 1984). The model has been used in both deterministic and probabilistic forms. In the deterministic form, it is proposed that the true line lies within the band with probability 1·0, and thus never deviates outside it. In the probabilistic form, on the other hand, the band is compared to a standard deviation, or some average deviation from the true line. One might assume that a randomly chosen point on the observed line had a probability of 68% of lying within the band, by analogy to the percentage of the normal curve found within one standard deviation of the mean.

Some clarification is necessary in dealing with line errors. In principle, we are concerned with the differences between some observed line, represented as a sequence of points with intervening straight line segments, and a true line. The gross misfit between the two versions can be measured readily from the area contained between them, in other words the sum of the areas of the spurious or sliver polygons. To determine the mismatch for a single point is not as simple, however, since there is no obvious basis for selecting a point on the true line as representing the distorted version of some specific point on the observed line. Most researchers in this field have made a suitable but essentially arbitrary decision, for example that the corresponding point on the true line can be found by drawing a line from the observed point which is perpendicular to the observed line (Keefer et al. 1988). Using this rule, we can measure the linear displacement error of any selected point, or compute the average displacement along the line.

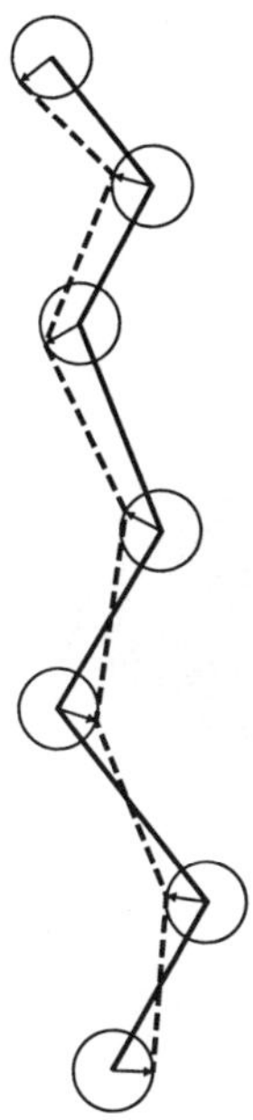

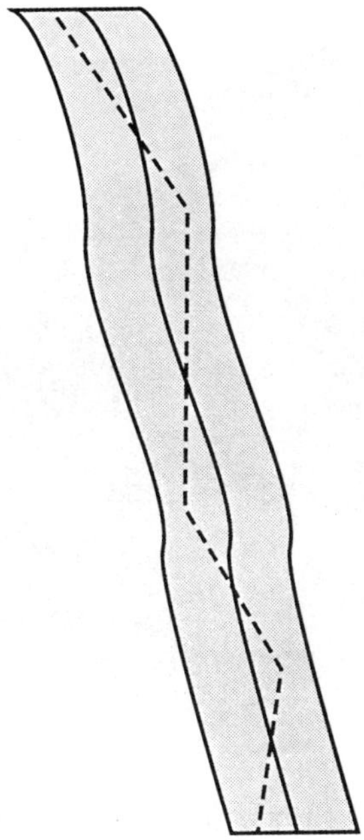

FIGURE 5 *left* Model of line distortion based on independent uncertainties in each digitized point
FIGURE 6 *right* The Perkal epsilon band

Although the Perkal band is a useful concept in describing errors in the representation of complex objects and in adapting GIS processes to uncertain data, it falls short as a stochastic process model of error. An acceptable model would allow the simulation of distorted versions of a line or polygon, which the Perkal band does not, and would provide the basis for a comprehensive analysis of error.

Alternatives to the Perkal Band

Goodchild and Dubuc (1987) have described a stochastic process which might be used to simulate the effects of error in soil, vegetation or forest cover maps. Consider the surface shown in Figure 7, which has the following properties. It is continuous, the elevation at every grid cell being a small displacement up or down from its neighbours, and therefore strongly spatially autocorrelated. It is also random, the surface having been generated by one of a number of stochastic processes known to produce random fields of this nature. In fact the surface was generated by the fractional Brownian process (Mandelbrot 1982), although other methods such as moving bands, spatially autoregressive and Markov processes have been described (for review see Haining et al. 1983). The

FIGURE 7 A spatially autocorrelated surface (random field) generated by the fractional Brownian process

fractional Brownian process has the property of self-affinity, that is, in principle any part of the surface if suitably expanded would be indistinguishable in its statistical properties from the surface as a whole.

Suppose now that the surface is classified by assigning each pixel or grid cell a colour or class based on its elevation, using a set of elevation classes or ranges. The result would be a standard contour map, in which the image is divided into a number of bands, each corresponding to a range of elevation, with the rule that classes may not be adjacent in the image unless the corresponding ranges of elevation are adjacent. The boundaries of the polygons so formed may not intersect. Finally, boundaries can be distorted to simulate the effects of digitizing errors by adding a suitably autocorrelated surface of small relief to the existing one and recontouring.

To simulate the appearance of a soil or vegetation map, Goodchild and Dubuc (1987) took two independently generated surfaces and passed each pixel through a simple classifier as in the example shown in Figure 8. This process is analogous to the control of ecological zones by temperature and precipitation in the model of Holdridge et al. (1971). The resulting map has the required properties; boundaries intersect in nodes, and have a degree of fragmentation and boundary wiggliness which depends on the ruggedness of the underlying surfaces. Distorted versions of the map can be produced by adding autocorrelated surfaces to one or both of the underlying surfaces, as in the two versions of the same map crossclassified

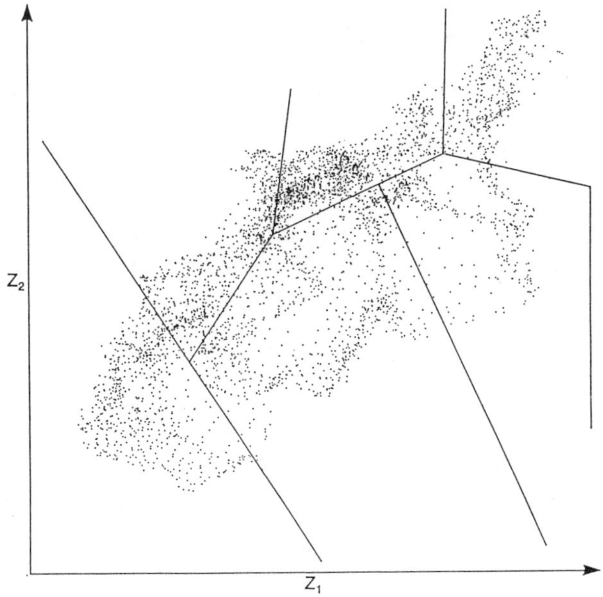

FIGURE 8 Use of a simple classifier to convert continuous surfaces to simulated area objects

in Table 3. Moreover the degree of distortion generated in each line will be a function of the classes which the line separates. In the simulated map shown in Figure 9 the boundaries have been smoothed by using a spline function, simulating the tendency of interpreters to smooth out lines of discontinuity in images.

The model is useful as a simulation of the distortion process which can be used to study the effects of uncertainty on GIS operations. However the number of parameters in the model, which include the number and properties of the underlying surfaces, the pixel size, the spline function and the classifier, argues against its calibration against real data and therefore against its use in characterizing the error in real datasets. In practice, GIS source maps tend to be highly abstracted, and to contain very little information on which a practical model of error might be based. It is only by returning to the original sources of the map's information that we can learn more about the nature of map error. In fact the map itself may be a barrier to better error modelling, by interposing a high level of abstraction between the raw data and the spatial database. Error modelling might be much easier if the raw imagery and ground surveys could be used to create a spatial database, and the abstract map obtained from the database by some objective process, rather than the reverse.

When the source of information is a remotely sensed image, a similar situation arises when pixels are classified and the result passed to a GIS for analysis, as this process similarly discards information which might be

TABLE 3 Sample Crossclassification Matrix Simulated by Overlaying Two Distorted Versions of the Same Map

Class	*Class*								
	A	*B*	*C*	*D*	*E*	*F*	*G*	*H*	*I*
A	561	0	0	0	2	0	44	14	20
B	0	170	0	15	0	0	0	0	0
C	0	0	0	0	0	0	0	0	0
D	0	1	0	315	0	33	0	0	0
E	2	0	0	0	429	0	0	28	60
F	0	0	0	2	0	684	58	0	20
G	5	0	0	0	0	7	544	0	86
H	17	0	0	0	21	0	0	337	1
I	21	0	0	0	10	0	14	3	572

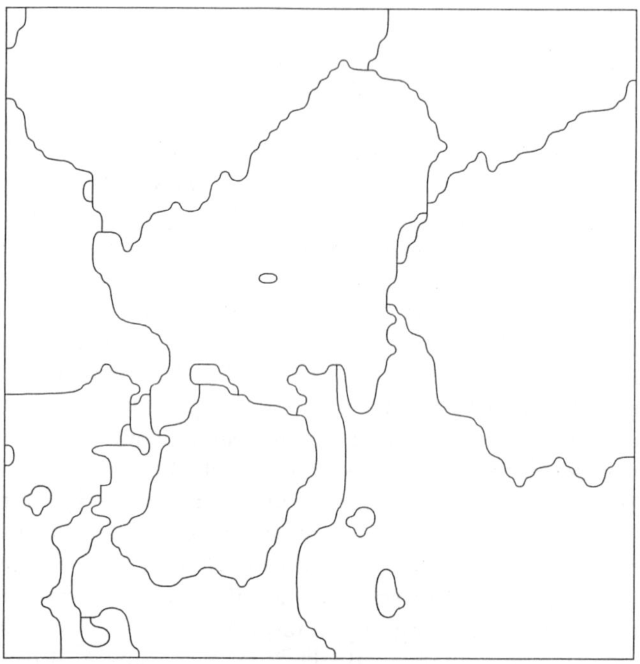

FIGURE 9 Simulation of area objects by the method of Goodchild and Dubuc (1987)

useful in modelling error. Many classification methods generate probabilities of class membership for each pixel, although usually only the most likely class is used. If the complete set of probabilities were passed to the GIS, then it would be possible to use these as the basis for uncertainty in

GIS products. Goodchild and Wang (1988) describe a stochastic process of object distortion based on this concept. In effect, both of these methods provide links between uncertainty in the continuous variation of a field, and uncertainty in the objects used to model the field in a spatial database for phenomena such as soils and vegetation.

Models for Dot and Pixel Counting

One of the traditional methods of measuring area from a map involves placing an array of grid cells or dots over the area to be measured, and counting. In principle this process is similar to that of obtaining the area of a patch from an image by counting the pixels which have been classified as belonging to or forming the patch. The accuracy of the area estimate clearly depends directly on the density of dots or the pixel size, but so does the cost of the operation, so it would be useful to know the precise relationship in order to make an informed judgment about the optimum density or size.

The literature on this topic has been reviewed by Goodchild (1980). In essence two extreme cases have been analysed, although intermediates clearly exist. In the first the area consists of a small number of bounded patches, often a single patch; in the second, it is highly fragmented so that the average number of pixels or dots per fragment of area is of the order of one or two. We refer to these as cases A and B respectively. Case A was first analysed by Frolov and Maling (1969), and later by Goodchild (1980). The limits within which the true area is expected to lie 95% of the time, expressed as a percentage of the area being estimated, are given by

$$\pm\ 1{\cdot}03\ (kn)^{1/2}(SD)^{-3/4}$$

where n is the number of patches forming the area; k is a constant measuring the contortedness of the patch boundaries, as the ratio of the perimeter to 3·54 times the square root of area ($k=1$ for a circle); S is the estimated area; and D is the density of pixels per unit area.

The results for Case B follow directly from the standard deviation of the binomial distribution, since this case allows us to assume that each pixel is independently assigned. The limits within which the true area is expected to lie 95% of the time, again as a percentage of the area being estimated, are given by

$$\pm\ 1{\cdot}96\ (1-S/S_0)^{1/2}(S/D)^{1/2}$$

where S_0 is the total area containing the scattered patches.

Error rises with the 3/4 power of cell size in case A, but with the 1/2 power in case B. Thus a halving of cell size will produce a more rapid improvement in error in case A, other things being equal, because only cells which intersect the boundary of the patch generate error in case A, whereas all cells are potentially sources of error in case B, irrespective of cell size.

Error in Images

Images such as those derived from remote sensing are composed of pixels (picture elements) of uniform size and shape, with associated measures of reflected or emitted radiation. After classification, each pixel is assigned to one of a number of classes, often of land use or land cover. Unlike the cartographic case, there are no objects or features to be located, and accuracy is simply a function of the errors in the assignment of classes to each pixel. The standard method of measuring accuracy in a classified image is to compare the classes assigned to a sample of pixels to the true classes on the ground ('ground truth') and express the result in the form of a table, the rows representing the assigned class and the columns the ground truth. Pixels which fall on the diagonal of the table are correctly classified; pixels which fall off the diagonal in a column are termed 'errors of omission' since the true value is omitted from the assigned class; and pixels which fall off the diagonal in a row are termed 'errors of commission' since they appear as false occurrences of a given class in the data. The contents of the table can be summarized in statistics such as the percentage of cells correctly classified, or using Cohen's Kappa statistic, which allows for correct classification by chance (see van Genderen and Lock 1977, van Genderen et al.1978, Mead and Szajgin 1982, Congalton et al. 1983, Greenland et al. 1985, Rosenfield 1986; Rosenfield and Fitzpatrick-Lins 1986).

PRACTITIONER'S SUMMARY

The measurement of data quality in cartographic products is dealt with in numerous publications and standards, many of them referenced in this chapter. Perhaps the most comprehensive view of cartographic data standards is that developed in the US by the Digital Cartographic Data Standards Task Force (DCDSTF 1988), which provides a five-fold model for describing data quality, although it sets no precise numerical thresholds in any of the five categories and defines no standard measurement procedures.

This chapter has taken the view that measurement of data quality for the digital databases now being developed to support GIS is substantially different from measurement of the accuracy of cartographic features. In fact the two questions 'Is this feature correct?' and 'What is at this place?'

require entirely different strategies. Some of these already exist, particularly in the feature-oriented domain, because they have been developed in support of traditional map-making. Others, particularly in the GIS-oriented domain, are still very limited and inadequate. Despite recent advances, the GISs of 1990 still regard the contents of the database as perfectly accurate, and make no attempt to estimate the uncertainties which are inevitably present in their products.

To summarize the state-of-the-art, we have good techniques for describing and measuring:

- the accuracy of location of a single point;
- the accuracy of a single measured attribute;
- the probability that a point at a randomly chosen location anywhere on a map has been misclassified;
- the effects of digitizing error on measures of length and area;
- the propagation of errors in raster-based area class maps through GIS operations such as overlay;
- the uncertainty in measures of area derived from dot counting.

However many of these are still in the domain of research literature, and we are in a very early stage of implementing them in standard practice.

From a practical point of view, the coming decade will see a greater awareness of the problems caused by uncertainty in the manipulation and analysis of geographical data, driven largely by the increasing availability of GIS. There will be increasing awareness also of the distinction between accuracy of cartographic features, and accuracy of GIS databases and products. We will see the emergence of systems equipped with tools for tracking uncertainties and the lineage of data through a complex series of operations, allowing the user to visualize and interact directly with such information through graphic interfaces. And finally, we will see increasing pressure on the providers of data to supply information on data quality.

PROJECTIVE SUMMARY

As we have seen, techniques for dealing with uncertainty in spatial data are much better developed in some areas than others. Uncertainties in point locations are comparatively easy to deal with, and a significant proportion of geodetic science is devoted to dealing with the issue of accuracy in control networks. Substantial progress has been made in modelling the error introduced by digitizing, and in predicting its effects on the measures of perimeter and area of objects (Chrisman and Yandell

1988, Griffith 1989). But the more general problem of modelling the accuracy of spatial data with respect to ground truth remains. In part this is because of the complexity of the processes involved, and in part because many ground truth definitions are themselves uncertain. But in addition, we have seen how the representation of spatial variation as objects acts in many cases to make error modelling more difficult, simply because of the nature of the object data model. Both of the models of error in area-class maps presented above are defined first in terms of continuous variation, using the raster data model, and then used as the basis for deriving uncertainty in objects. If the accuracy issue leads us to question the use of traditional data models then any progress will certainly be slow.

The accuracy of spatial databases is the topic of the first research initiative of the National Center for Geographic Information and Analysis (NCGIA), which began with a meeting of specialists in Montecito, CA, in December 1988. The collection of papers from that meeting has been published (Goodchild and Gopal 1989) and provides a broad overview of the accuracy issue in spatial data handling generally. The meeting identified an eleven-point research agenda:

- data structures and models;
- models of error and distortion;
- error propagation;
- product uncertainty and sensitivity;
- risk analysis;
- accuracy concerns among users and agencies;
- experimentation and measurement;
- error reduction methods;
- interpolation and surface modelling;
- aggregation, disaggregation and modifiable areal units;
- regularity and stability.

If accuracy is a problem in spatial data handling because of the high precision of the handling system, then one rational approach would be to reduce precision to match accuracy. In effect this is what happens in a raster system. Dutton (1984, 1989; see also Goodchild and Yang 1989) makes this point, and proposes a tesseral scheme for global data that would replace common coordinate referencing with a finite-resolution hierarchical key.

There has been substantial progress in recent years in understanding the propagation of uncertainty that occurs in spatial data handling. Newcomer and Szagin (1984) and Veregin (1989*c*) have discussed error propagation during simple Boolean overlay; Lodwick (1989), Lodwick and Monson (1988) and Heuvelink et al. (1989) have analysed the sensitivity of weighted overlay results to uncertainty in the input layers and weights; and Arbia and Haining (1990) have analysed a general model of uncertainty in raster data. However, overlay is perhaps the most simple of the primitive spatial data handling operations. Much more research is needed on the effects of buffering (dilating) uncertain objects, and of changing from one data model to another, eg raster/vector conversion.

If traditional map data models tend to give a misleading impression of the reliability of data, it is not surprising that many cartographic products fail to make the user aware of uncertainties. The move to standards of data quality (Moellering, Chapter 7) is a significant step towards greater awareness of the reliability of spatial data, and includes concepts of lineage, consistency and completeness as well as the more statistical issues discussed in this chapter. It is particularly important as GIS users continue to apply spatial data to purposes for which they may not have been designed.

This chapter has covered a range of material broadly connected with the issue of accuracy in spatial databases. As we noted at the outset, there are good reasons for believing that GIS technology will enhance our sensitivity to the nature of spatial data, and will increase the need to develop comprehensive and precise models of its errors and distortions. This enhanced understanding is likely to come not from study of the abstract objects which populate our maps and databases, but from the processes which created spatial variation in the first place and by which that variation was interpreted and modelled. In this sense GIS technology will help to fill many of the gaps which currently exist in our understanding of spatial differentiation and the processes which produce it.

It seems unlikely that the ideal accuracy-sensitive GIS which was outlined earlier in this chapter will ever become a reality. Its analogue in simple, aspatial measurement, the statistical analysis system which tracks uncertainty as an attribute of each of its data elements and through each of its procedures, is also readily justified but equally far from implementation at this time. On the other hand standard errors of estimate and confidence intervals are a common product of statistical systems, and it is not difficult to imagine similar products from GIS procedures.

One of the results of the information age is an erosion of the role of the central data collection agencies, and society has not yet begun to come to grips with a growing need to reorganize the entire system of public data provision (Haywood, Chapter 4). It is now possible to determine position at low cost and at any location using satellite receivers, and to do so with an accuracy which exceeds that of the USGS 1 : 24 000 map series or other readily available and authoritative sources. Population counts from local agencies are increasingly being used as alternatives to those of the Bureau of the Census. Accuracy is now increasingly seen as an attribute of spatial information, and as a component of its value to decision makers. In some areas its associated costs can be dramatic, when disputes based on spatial information involve litigation. The complex relationships between accuracy and cost in the GIS arena will be a challenging area for research for many years to come.

REFERENCES AND BIBLIOGRAPHY

American Society of Photogrammetry (Committee for Specifications and Standards, Professional Practice Division) (1985) Accuracy specification for large-scale line maps. *Photogrammetric Engineering and Remote Sensing* **51** (2): 195–9

Amrhein C, Griffith D A (1987) GIS, spatial statistics and statistical quality control. Paper presented at *IGIS* '87, Washington DC

Arbia G, Haining R P (1990) Error propagation through map operations. (Unpublished)

Aronoff S (1982) Classification accuracy: A user approach. *Photogrammetric Engineering and Remote Sensing* **48** (8): 1299–1307

Aronoff S (1985) The minimum accuracy value as an index of classification accuracy. *Photogrammetric Engineering and Remote Sensing* **51** (1): 99–111

Barrett J P, Philbrook J S (1970) Dot grid area estimates: Precision by repeated trials. *Journal of Forestry* **68** (3): 149–51

Bauer M E, Hixson M M, Davis B J, Etheridge J B (1978) Area estimation of crops by digital analysis of Landsat data. *Photogrammetric Engineering and Remote Sensing* **44** (8): 1033–43

Baugh I D H, Boreham J R (1976) Measuring the coastline from maps: A study of the Scottish mainland. *The Cartographic Journal* **13** (2): 167–71

Beard M K (1989) Use error: The neglected error component. In: *Proceedings, AutoCarto* 9. Falls Church, VA, ASPRS/ACSM: 808–17

Bedard Y (1987) Uncertainties in land information systems databases. In: *Proceedings, AutoCarto* 8. Falls Church, VA, ASPRS/ACSM: 175–84

Blakemore M (1984) Generalization and error in spatial databases. *Cartographica* **21**: 131–9

Blakemore M (1985) High or low resolution? Conflicts of accuracy, cost, quality and application in computer mapping. *Computers and Geosciences* **11** (3): 345–8

Blakney W G G (1968) Accuracy standards for topographic mapping. *Photogrammetric Engineering and Remote Sensing* **34** (10): 1040–2

Bracken I, Martin D (1989) The generation of spatial population distributions from census centroid data. *Environment and Planning A* **21** (8): 537–44

Burrough P A (1986) *Principles of Geographical Information Systems for Land Resources Assessment*. Oxford, UK, Clarendon Press

Campbell J B (1977) Variation of selected properties across a soil boundary. *Soil Science Society of America Journal* **41** (3): 578–82

Campbell J B, Edmonds W J (1984) The missing geographic dimension to soil taxonomy. *Annals of the Association of American Geographers* **74** (1): 83–97

Card D H (1982) Using known map category marginal frequencies to improve estimates of thematic map accuracy. *Photogrammetric Engineering and Remote Sensing* **48** (3): 431–9

Carter J R (1989) Relative errors identified in USGS gridded DEMs. In: *Proceedings, AutoCarto* 9. Falls Church, VA, ASPRS/ACSM: 255–65

Chrisman N R (1982) Methods of spatial analysis based on errors on categorical maps. Unpublished PhD Dissertation, Department of Geography, University of Bristol

Chrisman N R (1983) The role of quality information in the long-term functioning of a geographic information system. *Cartographica* **21** (2/3): 79–87

Chrisman N R (1987) The accuracy of map overlays: A reassessment. *Landscape and Urban Planning* **14**: 427–39

Chrisman N R (1989) Error in categorical maps: Testing versus simulation. In: *Proceedings, AutoCarto* 9. Falls Church, VA, ASPRS/ACSM: 521–9

Chrisman N R and Yandell B (1988) A model for the variance in area. *Surveying and Mapping* **48**: 241–6

Congalton R G (1988*a*) Using spatial autocorrelation analysis to explore the errors in maps generated from remotely sensed data. *Photogrammetric Engineering and Remote Sensing* **54** (5): 587–92

Congalton R G (1988*b*) A comparison of sampling schemes used in generating error matrices for assessing the accuracy of maps generated from remotely sensed data. *Photogrammetric Engineering and Remote Sensing* **54** (5): 593–600

Congalton R C, Oderwald R G, Mead R A (1983) Assessing Landsat classification accuracy using discrete multivariate analysis statistical techniques. *Photogrammetric Engineering and Remote Sensing* **49** (12): 1671–8

Crapper P F (1980) Errors incurred in estimating an area of uniform land cover using Landsat. *Photogrammetric Engineering and Remote Sensing* **46** (10): 1295–1301

Crapper P F, Walker P A, Manninga P M (1986) Theoretical prediction of the effect of aggregation on grid cell data sets. *Geoprocessing* **3** (2): 155–66

Curran P J, Williamson H D (1985) The accuracy of ground data used in remote sensing investigations. *International Journal of Remote Sensing* **6** (10): 1637–51

Dahlberg R E (1986) Combining data from different sources. *Surveying and Mapping* **46** (2): 141–6

Digital Cartographic Data Standards Task Force (1988) The proposed standard for digital cartographic data. *The American Cartographer* **15** (1): 1–140

Dutton G (1984) Geodesic modeling of planetary relief. *Cartographica* **21**: 188–207

Dutton G (1989) Modeling locational uncertainty via hierarchical tesselation. In: Goodchild M F and Gopal S (ed.) *Accuracy of Spatial Databases*. New York, Taylor and Francis: 125–40

Flowerdew R, Green M (1988) Statistical methods for inference between incompatible zoning systems. In: Goodchild M F and Gopal S (ed.) *Accuracy of Spatial Databases*. New York, Taylor and Francis: 239–48

Franklin W R (1984) Cartographic errors symptomatic of underlying algebra problems. In: *Proceedings, International Symposium on Spatial Data Handling, Zurich*. University of Zurich: 190–208

Franklin W R, Wu P Y F (1987) A polygon overlay system in PROLOG. In: *Proceedings, AutoCarto* 8. Falls Church, VA, ASPRS/ACSM: 97–106

Frolov Y S, Maling D H (1969) The accuracy of area measurements by point counting techniques. *Cartographic Journal* **6**: 21–35 van Genderen J L, Lock B F (1977) Testing land-use map accuracy. *Photogrammetric Engineering and Remote Sensing* **43** (9): 1135–7

van Genderen J L, Lock B F, Vass P A (1978) Remote sensing: Statistical testing of thematic map accuracy. *Remote Sensing of Environment* **7** (1): 3–14

Goodchild M F (1979) Effects of generalization in geographical data encoding. In: Freeman H and Pieroni G G (ed.) *Map Data Processing*. New York, Academic Press: 191–206

Goodchild M F (1980) Fractals and the accuracy of geographical measures. *Mathematical Geology* **12**: 85–98

Goodchild M F (1989) Modeling error in objects and fields. In: Goodchild M F and Gopal S (ed.) *Accuracy of Spatial Databases*. New York, Taylor and Francis: 107–14

Goodchild M F, Dubuc O (1987) A model of error for choropleth maps, with applications to geographic information systems. In: *Proceedings, AutoCarto* 8. Falls Church, VA, ASPRS/ACSM: 165–74

Goodchild, M F, Gopal, S (1989) *Accuracy of Spatial Databases*. New York, Taylor and Francis

Goodchild M F, Lam N S (1980) Areal interpolation: A variant of the traditional spatial problem. *Geoprocessing* **1**: 297–312

Goodchild M F, Wang M H (1988) Modeling error in raster-based spatial data. In: *Proceedings, Third International Symposium on Spatial Data Handling, Sydney*. Columbus, OH, IGU Commission on Geographical Data Sensing and Processing: 97–106

Goodchild M F, Wang M H (1989) Modeling errors for remotely sensed data input to GIS. In: *Proceedings, AutoCarto* 9. Falls Church, VA, ASPRS/ACSM: 530–7

Goodchild M F, Yang S (1989) A hierarchical spatial data structure for global geographic information systems. Technical Report 89–5, National Center for Geographic Information and Analysis, University of California, Santa Barbara, CA

Greenland A, Socher R M, Thompson M R (1985) Statistical evaluation of accuracy for digital cartographic data bases. In: *Proceedings, AutoCarto* 7. Falls Church, VA, ASPRS/ACSM: 212–21

Griffith D A (1989) Distance calculations and errors in geographic databases. In: Goodchild M F and Gopal S (ed.) *Accuracy of Spatial Databases*. New York, Taylor and Francis: 81–90

Haining R P, Griffith D A, Bennett R J (1983) Simulating two-dimensional autocorrelated surfaces. *Geographical Analysis* **15**: 247–53

Hakanson L (1978) The length of closed geomorphic lines. *Mathematical Geology* **10** (2): 141–67

Hay A M (1979) Sampling designs to test land-use map accuracy. *Photogrammetric Engineering and Remote Sensing* **45** (4): 529–33

Hay A M (1988) The derivation of global estimates from a confusion matrix. *International Journal of Remote Sensing* **9** (8): 1395–8

Heuvelink G B M, Burrough P A, Stein A (1989) Propagation of errors in spatial modelling with GIS. *International Journal of Geographical Information Systems* **3** (3): 303–22

Holdridge L R, Grenke W C, Hathaway W H, Liang T, Tosi J A Jr. (1971) *Forest Environments in Tropical Life Zones: A Pilot Study*. Oxford, Pergamon

Honeycutt D M (1986) Epsilon, generalization and probability in spatial data bases (Unpublished)

Hudson D, Ramm C W (1987) Correct formulation of the kappa coefficient of agreement. *Photogrammetric Engineering and Remote Sensing* **53** (4): 421–2

Keefer B J, Smith J L, Gregoire T G (1988) Simulating manual digitizing error with statistical models. In: *Proceedings, GIS/LIS* '88. Falls Church, VA, ASPRS/ACSM: 475–83

Lee Y C (1985) Comparison of planimetric and height accuracy of digital maps. *Surveying and Mapping* **45** (4): 333–40

Lloyd P R (1976) Quantisation error in area measurement. *The Cartographic Journal* **15** (1): 22–5

Lodwick W A (1989) Developing confidence limits on errors of suitability analyses in geographical information systems. In: Goodchild M F and Gopal S (ed.) *Accuracy of Spatial Databases*. New York, Taylor and Francis: 69–78

Lodwick W A, Monson W (1988) Sensitivity analysis in geographic information systems. Department of Mathematics, University of Colorado at Denver, Paper RP886.

MacDougall E B (1975) The accuracy of map overlays. *Landscape Planning* **2** (1): 23–30

MacEachren A M (1985) Accuracy of thematic maps: implications of choropleth symbolization. *Cartographica* **22** (1): 38–58

Maling D H (1968) How long is a piece of string? *The Cartographic Journal* **5** (2): 147–56

Maling D H (1989) *Measurement from Maps: Principles and Methods of Cartometry*. New York, Pergamon

Mandelbrot B B (1967) How long is the coast of Britain? *Science* **156**: 636–8

Mandelbrot B B (1982) *The Fractal Geometry of Nature*. San Francisco, Freeman

Mark D M, Csillag F (1989) The nature of boundaries on area–class maps. *Cartographica* **21**: 65–78

Mead R A, Szajgin J (1982) Landsat classification accuracy assessment procedures. *Photogrammetric Engineering and Remote Sensing* **48** (1): 139–41

Merchant D C (1987) Spatial accuracy standards for large scale topographic maps. *Photogrammetric Engineering and Remote Sensing* **53** (7): 958–61

Moellering H (ed.) (1987) *A Draft Proposed Standard for Digital Cartographic Data*. Report No. 8, National Committee for Digital Cartographic Data Standards, Columbus.

Muller J-C (1977) Map gridding and cartographic errors: A recurrent argument. *Canadian Cartographer* **14** (2): 152–67

Muller J-C (1987) The concept of error in cartography. *Cartographica* **24** (2): 1–15

Newcomer J A, Szajgin J (1984) Accumulation of thematic map errors in digital overlay analysis. *The American Cartographer* **11** (1): 58–62

Openshaw S (1977*a*) A geographical solution to scale and aggregation problems in region-building, partitioning and spatial modelling. *Institute of British Geographers, Transactions* **2** (NS): 459–72

Openshaw S (1977*b*) Optimal zoning systems for spatial interaction models. *Environment and Planning A* **9**: 169–84

Openshaw S (1978) An empirical study of some zone-design criteria. *Environment and Planning A* **10**: 781–94

Openshaw S, Taylor P J (1979) A million or so correlation coefficients: Three experiments on the modifiable areal unit problem. In: Wrigley N (ed.) *Statistical Applications in the Spatial Sciences*. London, Pion

Openshaw S, Taylor P J (1981) The modifiable areal unit problem. In: Wrigley N and Bennett R J (ed.) *Quantitative Geography: A British View*. London, Routledge

Perkal J (1956) On epsilon length. *Bulletin de l'Academie Polonaise des Sciences* **4**: 399–403

Perkal J (1966) On the Length of Empirical Curves. Discussion Paper No. 10, Michigan Inter-University Community of Mathematical Geographers, Ann Arbor

Robinson J E (1973) Frequency analysis, sampling, and errors in spatial data. In: Davis J C and McCullagh M J (ed.) *Display and Analysis of Spatial Data*. London, Wiley: 78–95

Rosenfield G H (1971) On map accuracy specifications: Part II. Horizontal accuracy of topographic maps. *Surveying and Mapping* **31** (1): 60–4

Rosenfield G H (1986) Analysis of thematic map classification error matrices. *Photogrammetric Engineering and Remote Sensing* **52** (5): 681–6

Rosenfield G H, Fitzpatrick-Lins K (1986) A coefficient of agreement as a measure of thematic map accuracy. *Photogrammetric Engineering and Remote Sensing* **52** (2): 223–7

Smith G R, Honeycutt D M (1987) Geographic data uncertainty, decision making and the value of information. In: *Proceedings, GIS '87, San Francisco*: 300–12

Stoms D (1987) Reasoning with uncertainty in intelligent geographic information systems. In: *Proceedings, GIS '87, San Francisco*: 693–700

Story M, Congalton R G (1986) Accuracy assessment: A user's perspective. *Photogrammetric Engineering and Remote Sensing* **52** (3): 397–9

Switzer P (1975) Estimation of the accuracy of qualitative maps. In: Davis J C and McCullagh M J (ed.) *Display and Analysis of Spatial Data*. London, Wiley: 1–13

Thompson M M (1971) On map accuracy specifications: Part I. Practical interpretation of map-accuracy specifications. *Surveying and Mapping* **31** (1): 57–60

Tobler W R (1979) Smooth pycnophylactic interpolation for geographical regions. *Journal of the American Statistical Association* **74**: 519–30

Veregin H (1989*a*) Accuracy of spatial databases: Annotated bibliography. Technical Report 89–9, National Center for Geographic Information and Analysis, University of California, Santa Barbara

Veregin H (1989*b*) A taxonomy of error in spatial databases. Technical Report 89–12, National Center for Geographic Information and Analysis, University of California, Santa Barbara

Veregin H (1989*c*) Error modeling for the map overlay operation. In: Goodchild M F and Gopal S (ed.) *Accuracy of Spatial Databases*. New York, Taylor and Francis: 3–18

Veregin H (1989*d*) A review of error models for vector to raster conversion. *The Operational Geographer* **7** (1): 11–15

Vitek J D, Richards D G (1978) Incorporating inherent map error into flood-hazard analysis. *Professional Geographer* **30** (2): 168–73

Walsh S J, Lightfoot D R, Butler D R (1987) Recognition and assessment of error in geographic information systems. *Photogrammetric Engineering and Remote Sensing* **53** (10): 1423–30

Webster R, Beckett P H T (1968) Quality and usefulness of soil maps. *Nature* **219**: 680–2

Wehde M (1982) Grid cell size in relation to errors in maps and inventories produced by computerized map processing. *Photogrammetric Engineering and Remote Sensing* **48** (8): 1289–98

Yoeli P (1984) Error bands of topographical contours with computer and plotter (program KOPPE). *Geoprocessing* **2** (3): 287–97

Chapter 7

CARTOGRAPHIC DATABASE EXCHANGE STANDARDS

Harold Moellering
Department of Geography, 103 BK, Ohio State University, Columbus, Ohio 43210, USA

INTRODUCTION

During the last decade many national efforts have been initiated to develop cartographic database exchange standards. At this point in time, most of these efforts are still actively in progress. Work on such standards is underway in such countries as Austria, Germany, France, the UK, Finland, Canada, the USA, South Africa and Japan among others. International organizations such as the International Hydrographic Organization, the Digital Geographic Information Working Group (recently spun off from NATO), Comité Européen des Responsables de la Cartographie Officielle (CERCO), and the Digital Chart of the World group (DCW) are also active. A summary of these international activities is scheduled to appear in Moellering (1991).

The USA has had one of the oldest and most comprehensive efforts to develop such database transfer standards. The work began in the early 1980s with the work of the ACSM National Committee for Digital Cartographic Data Standards in cooperation with the US Geological Survey. The first work of this effort was a thrust to define the specific nature and goals of this standards effort (Moellering 1982). Shortly thereafter the Standards Working Group of the Federal Coordinating Committee of Digital Cartography began to work in this area. In 1987 the National Committee issued a Draft Proposed Standard (Moellering 1987). About the same time the Federal Committee also issued a set of proposals (TeSelle, 1986). This work was melded together where the National Committee Draft Proposed Standard served as the base document for the standard and a number of features from the Federal Committee proposals were incorporated into it. This resulted in the Proposed Standard published in the January issue of the *American Cartographer*, (DCDSTF 1988). This proposed standard was once again polished, empirically field tested, and circulated for comments. The result from all of these efforts is the US Spatial Database Transfer Standard (SDTS)(SDTSTRB 1990). During this nearly ten-year period the scientific and technical progress of the effort was hindered or occasionally stymied on many occasions by the weakness of the theoretical and conceptual basis for spatial databases and their transfer mechanisms. Time after time these groups encountered 'gaps in our knowledge' which limited the work.

This chapter discusses the research needs for spatial data transfer and its surrounding elements that have been identified during the course of this decade of effort. Because of the background of the author the discussion will focus primarily on the American effort, but the reader should understand that many of the research needs identified in this discussion also apply to efforts in progress elsewhere in the world. The following sections of this chapter will initially focus on broader conceptual research needs and then on the more specific areas of spatial objects, data quality, entities, the exchange mechanism, and finally on research needs that relate to international comparisons of these developing standards. A reasonable knowledge of SDTS is assumed in this discussion.

THE NEED TO EXPAND BASIC CONCEPTS

During the decade of this effort the field has witnessed a change from simple 'computer' cartography to a more conceptual and sophisticated orientation towards analytical cartography. The fundamental concept is to understand that such database transfers are exchanges of virtual maps. The concepts of real and virtual maps was first developed by Moellering (1980) and are a significant conceptual extension of the transformational concepts in cartography originally proposed by Tobler(1979). A virtual map type 3 cannot be directly viewed and does not have a permanent tangible reality. An excellent example of such a virtual map is a digital spatial database. An exchange of such a database from one system to another is a virtual 3 to virtual 3 transformation. What is really being exchanged is not a graphic image as many people think, but the data that underlie the cartographic image. Nyerges(1980) has defined this non-graphic domain that contains a very rich set of spatial and attribute relationships as the deep structure. The cartographic image is defined as surface structure. Even today these crucial concepts are not fully understood or appreciated in cartography. It is clear that these concepts must be explored in more detail, their implications for spatial database exchanges and cartography in general must be recognized and expanded.

A related topic is the concern to develop a deeper understanding of cartographic data levels (Nyerges 1980) as they relate to the syntax and semantics of the spatial data transfer process. Nyerges has defined six cartographic data levels as the data reality: information structure, canonical structure, data structure, storage structure and the machine encoding. These levels relate in certain ways to the syntax of the data transfer, the actual structures captured in the database, but also to the semantics, which is the meaning of what is being transferred. Although this topic has been explored by Nyerges(1984) and by Pascoe and Penny (1990), it is clear that much more research is necessary to truly understand this process in conceptual depth. To date most standards efforts have concentrated on encoding spatial data in some sort of transfer structure or mechanism, which is the syntax. But not as much effort has been devoted to understanding the semantics or meaning of the spatial data being exchanged. It appears that the syntax part of the exchange process is the easy part and that the semantics part of the exchange process is much more difficult. More very basic conceptual research is needed to deepen our understanding of this process.

Over the years many kinds of spatial data models have been proposed, tested and implemented. See, for example, Peuquet (1984). We now find ourselves with three fundamental families of data structures: vector, raster, and relational. Currently, in most cases, these families must be treated separately in a spatial data transfer, but current needs suggest that concepts must be developed to integrate these sorts of data structures together. One pressing need is to efficiently handle raster and vector data together. Although such raster/vector hybrid structures have been proposed (Peuquet 1983, Abel and Wilson 1990) and empirically tested (Ross 1987), much more research is required. Bouille (1978, 1984) has proposed the hypergraph data structure. Clearly many other possibilities must be investigated.

In the past few years richer spatial database concepts have begun to emerge. The notion of object-oriented programming was originally developed in computer science (see Danforth and Tomlinson 1988) and now are beginning to be applied to spatial data (Egenhofer and Frank 1989, Kemp 1990). Although the SDTS recognizes and contains capabilities for the transfer of object-oriented databases, more research is required in the spatial data community to really work out the effective use of such structures. Related to this development is the notion of image databases. (See, for example, Grosky and Mehrotra 1989.) This offers much potential for cartographic uses, but has been little discussed in literature. Recent discussions in the literature have also suggested the possibility of hypermaps, as discussed by Laurini and Milleret-Raffort (1989), which are really hyperdata structures. The hypermap concept seems to offer much potential for capturing and processing spatial data.

What is needed is much more research to investigate the efficient transfer of such databases.

As one sorts through the research work that has been done on spatial databases and spatial data structures it is easy to identify many pieces of quality research that have been done. It makes one wonder if these individual pieces of spatial data research really fit into some sort of larger whole. Therefore, one question that arises is whether there exists a general theory of spatial data similar to that of a general theory of relativity. Such a possibility must be investigated. What follows are comments that are much more specific to the development of SDTS.

RESEARCH ON SPATIAL OBJECTS

At present the understanding of spatial objects and manipulations on them is fairly reasonable, otherwise transfer standards could not have been developed at all. But there are still a number of gaps in our knowledge which should be filled. This discussion is an expansion of ideas discussed in Moellering (1987*b*).

Object Specification

The work on the specification of linear parametric curves still needs more work. This is because there are so many ways to specify parametric curves. The earliest versions of what developed into SDTS covered the general concept of parametric curves with the geometric arc primitive. However the later specification of the number of families of mathematical formulations that should be implemented created the problem. Earlier versions of SDTS specified about 15 formulations. However, as it became clear that the parameter values for a particular curve change with a change in projection, the situation became more clouded. The current strategy is to specify the half dozen formulations that are most used in the spatial sciences, and then to specify them in a rather general form. This approach meets the requirement but more research is required.

One of the major gaps in our knowledge is a good understanding of the most efficient ways to specify primitives for $2\frac{1}{2}$- and 3-dimensional objects. Although the 0-, 1- and 2-dimensional object primitives are fairly well understood, these are not. At the moment this gap is finessed by decomposing $2\frac{1}{2}$- and 3-dimensional spatial objects into their lower level dimensional primitives. They can then be transferred in this manner, but such a transfer could be complicated if the topological relationships are rich. What is desired is a compact and efficient set of such spatial primitives that cover all cases where such primitives are used. This task is complicated by the fact that there are many continuous and piecewise approaches to specifying surfaces and developing a set of such primitives that covers all of them will not be easy. Since a surface is based on variations of $Z = f(X,Y)$, a direct representation of such a surface in the form of primitives is clearly advantageous. The approach would have to take into account both discrete and continuous specifications of such surfaces. Since surface topology in cartography mostly utilizes what is available for 2-D objects, as discussed by Corbett(1979) and White (1979) that situation seems to be reasonable.

However, for 3-D primitives the situation is more complicated. Here not only are volumetric partitionings not clearly specified, but the general problem in cartography is not that well understood. Although there are things such as voxels and prisms that are really extensions from the 2-D primitives, notions of how to handle the irregular 3-D primitives is much less clear. The theory of irregular polytopes may provide some insight to help clarify the problem. It is clear that much more work on 3-D topology is necessary to help solve the problem. This question must be resolved for the geometry/topology primitives. It seems that point set topology could offer clues and needs to be investigated thoroughly (Frank 1990).

Another arena that needs more topological research is that of pixels and grid cells. Currently pixels and grid cells are specified as geometry-only objects. These current definitions ignore topology. Many operators can operate successfully in this environment because the neighbourhood topology is implicit due to the isotropic nature of the square cell structure. However, it is clear that vector data structures have greatly benefited from the integration of topological properties of connectivity and contiguity. It seems possible that a similar benefit might be feasible for pixels, grid cells, and the special case of raster data structures. The addition of topological properties to pixels and grid cells might facilitate the more efficient use of these objects in spatial data structures. It is even possible that the topological property could help to more efficiently solve the raster–vector conversion problem if a more efficient data structure homeomorphism could be identified. Such work could lead to additional hybrid raster–vector data structures that might work more efficiently than current vaster data structures as discussed above.

If one examines the current primitive object definitions in SDTS, two explicit classes of primitives will be noted. They are geometry only, and geometry and topology. Many in the field would think that those two classes would suffice. However, there is a third implicit class of primitive spatial objects that is mentioned in the specification, that of topology-only objects. Although most cartographers consider geometry and coordinates a required part of the mapping process, work by White and Griffin (1979) and others has shown that indeed one can work on cartographic problems that do not require coordinates. Work by Bouille on hypergraph data structures has used the concept of the hyperlink, an object primitive that does not use coordinates. Such hyperlinks are used to connect various kinds of objects and attributes in a deep structure environment where coordinates may not be necessary. There are other hyperspace situations where linkages are necessary in a database that do not involve coordinates. It is expected that such linkages will commonly occur in multimedia hypermaps where various kinds of logical linkages will be required to move smoothly between various expressions of spatial reality and its extracted representations. At the moment such primitives can be created by truncating the coordinates from the geometry/topology objects which is an implicit definition. Save for something like a hyperlink, explicit definitions have not yet evolved for such primitives. Currently so little is concretely understood of such topology-only primitives that names for specific primitives have not yet evolved.

The question of pictorial objects also arises. Currently pixels and grid cells have been explicitly defined and are used for image based systems. However, the expanding amount of work on pictorial databases suggests that this question should be re-examined to see if the two existing primitives are adequate for the cartography of the future. The idea of multimedia hypermaps suggests that direct imagery may be used in much more integrated ways with what we now think of as spatial databases. Having the tools for hyperlinkages inside such databases is one question, but having a full set of pictorial primitives is a different and very interesting question.

A totally different facet of the general question of spatial object specification is that of the temporal dimension. It is a given that the real world as cartographic reality exists in both time and space. In the last few centuries the spatial sciences focused almost exclusively on the spatial dimensions of the problem. It is only recently that work has commenced to explicitly design spatial data structures that can efficiently handle the temporal dimension. The efficient handling of the temporal dimension goes beyond the simple coding of it as a separate attribute, but as an additional orthogonal dimension. To date some work has been accomplished, but the true research question has not yet been clearly framed, much less analysed and answered. There is room for fruitful research in this area.

Manipulation and Use of Spatial Objects

The first question that arises from the use of spatial objects is the building of complex objects from the primitives. SDTS is designed to transfer databases between systems that have different hardware architecture and data models in them. It is clear that systems with different data models may well build complex objects in different ways with different combinations of primitives, even if they belong to the same family of data structures. An integral part of these complex objects is a potentially rich set of relationships that could exist between attributes of parts of an object, and attributes that apply to more than one part of the object substructure. The individual rules for the relationship of such a data model are usually specified in the canonical and data structure levels (Nyerges levels 3 and 4). However these specifics of the design of these relationship rules varied from data model to data model. Research is required to learn more about natural ways that these rules should be structured in order to facilitate the easy use and manipulation of these more complex objects. Revealing results from such investigations would help designers to build better and more efficient spatial databases which are more powerful, as well as to simplify the design and building of transfer metafile converters. The above effort would help to develop more efficient object-oriented databases that would facilitate such operations. Although this is only one class of data structure, the potential for the future seems bright.

One should consider developing more fully the family of spatial operators that work in 0-, 1-, 2- and 3-dimensions. Of these, the 1- and 2-dimensional operators are perhaps the best developed. Operators of 0-dimension are somewhat developed, especially if one assumes that they are point samples from a continuous surface. In some sense continuous surface operators can be considered as $2\frac{1}{2}$-dimensional operators while 3-dimensional operators in GIS and analytical cartography are practically unknown. However, some very interesting work has been done by Claire and Guptill (1982) and Tobler (1984) among others. What one would like to see is a periodic table of spatial operators that specifies them by dimension, scaling of the data, and perhaps other properties.

Another limitation of spatial operators one sees today is the fact that most analytical operators are data structure specific. The obvious example is one operation that is specified for some sort of vector data structure and must be reformulated to work properly in the raster domain. Ideally the desired environment for such operators would be one where analytical operators are data structure independent with perhaps some differences in the software implementation. It is possible that this cannot be accomplished until a more general theory of spatial data is developed, but it is an extremely interesting question to investigate. A significant improvement in this part of the theory would save software developers many millions of dollars and reduce the development time of commercially available systems.

Basic concepts of topology have been known and used in cartography for more than two decades. These concepts are well enough understood as topological applications in cartography for Corbett (1979) to have published a monograph on the subject. This work has not received sufficient recognition in the spatial sciences. Further work by White (1979) and White and Griffin (1979) has shown that topological operations are extremely helpful in a cartographic context. This work suggests that it may be possible to develop a set of topology-only operators for spatial objects and data structures. Bouille (1978, 1984) has shown that hypergraph linkages in data structures are not only useful, but preferred for certain kinds of data structures. His Hypergraph Based Data Structures (HBDS) and the principles that they are based on are very useful to spatial database designers.

To date most work in the spatial sciences has confined itself to the three physical dimensions of human space. However, it is possible to conceive of a data space that has many more dimensions. This is not a space directly specified by the data structure as shown above by Bouille, but a data space that has to do with the attributes of the spatial features that are being coded and analysed by spatial systems. Currently, it is easiest to analyse such a feature space in a parametric setting, but work in spatial cognition has shown that analysis of nonparametric space is also possible given certain constraints. Multidimensional classification of pixels in parametric space is fairly common in remote sensing. What is being suggested here is that similar kinds of analyses could be performed in spatial feature space. This space would have parametric measures, but work in nonparametric space is possible if little explored in analytical cartography. This sort of research would tie in quite well with work on hypermaps discussed above.

The last topic in this section deals with the development of a true spatial data query and manipulation language. Although many query languages have been developed in computer science over the last couple of decades, most treat the objects contained in them as being zero dimensional. Such query languages are generally useful in cartography and GIS except for special cases. Recently it has been realized that what is really needed is a true spatial data query language (Robinson, Chapter 5). Such a language should be developed as a general solution to the problem for object manipulation as well as retrieval and output. Given the current state of development of spatial database, object and operator concepts, this is a rather daunting challenge. However, it could also yield very fruitful results that would simplify the data transfer problem and the standards that are being designed to accomplish it.

QUALITY OF SPATIAL DATA

In any anticipated spatial database transfer it is very important to have an authoritative quality report on the data contained in the file. Such information will form a major component of the decision as to whether any particular database should be utilized for a particular purpose. Since different problems have different requirements for data accuracy, the quality report will be analysed carefully in making the decision as to whether the database is suitable for a particular use. The current SDTS quality report has five components: lineage, positional accuracy, attribute accuracy, logical consistency and completeness. Although the current specification provides for several levels of rigour in reporting such information, the information provided by the user is largely textual in nature. Although some of the information will be numerical, the overall quality report is textual. Here there is great room for improvement of the quality information provided to the user, although these possibilities must be balanced against the cost of providing such information.

The first major improvement to the process of data collection and database building is to adopt more rigorous methods of statistical quality control in that process itself (Goodchild, Chapter 6). Many of these methods could be adapted from the standard quality control (QC) literature. Some research will be required to decide which methods are practical and cost effective. At the present time many database elements are collected and assembled in a way that could be made more systematic and rigorous. If such QC methods are adopted for building spatial databases, the cost of reporting the SDTS required quality elements will probably be lower. Ballou and Tayi (1989) suggest a methodology for allocating resources to make such work more efficient. In any case, the quality of the data actually collected will probably be higher.

The current SDTS required quality elements give the data provider wide flexibility in how this quality information is reported. There is a great need to specify these quality levels in a uniform and consistent way so that the elements of the quality report can be compared quantitatively. This is different from product specific quality specifications where fixed numerical thresholds are required. Here the need is to specify a more rigorous range of standard quality control statistical measures for each reporting class. More research is required to implement this special application of quality control.

It is clear at the outset that topological structures are very perishable and easily upset. The current approach of reporting topological quality levels is still largely descriptive. What is needed is better quantitative methods to report such tests, both for understanding by the potential user and for comparison purposes. Such measures would test the placement of 0-D objects relative to related objects of other dimensions as well as to the well known characteristics of connectivity for linear networks and neighbourliness for adjoining areal objects. Chrisman (1982, 1989) has addressed a number of these problems associated with categorical maps and provided a general taxonomy of the problem. What is needed is to continue this work further into the analytical and statistical domain.

Lineage is also a very important component of the quality report. It is rare that an entire spatial database is coded and built at one time. Rather the general scenario is that valid elements of older databases are updated with new or replacement elements to form a new version of that database. Each database has a history or lineage which must be reported to the user. However lineage metadata on the database can be very detailed and potentially voluminous. Consider the legal requirements for maintaining lineage on nautical navigation aids in the waters of the USA. Each individual navigation aid can have a lengthy history and a wealth of information so that such a database on navigation features may be considered a hypermap. With possibilities for such a large volume of lineage information for what seems like a simple chart, it is clear that very efficient and systematized methods for storing such information are required. Although the case of nautical navigation aids has been examined, there are many other potential situations that will probably arise in the future. The improved handling of lineage metadata is an interesting applied research topic, and hopefully the proper tools will be ready when needed.

SPATIAL ENTITIES

The development of a national set of common entity definitions is a very labour intensive challenge. When the initial goal is accomplished, a continuing maintenance authority is required to keep the definitions of features and attributes up to date. A number of these requirements have been discussed by Rugg and Schmidt (1987) that relate to the implementation of SDTS. However there are several additional research topics that should be examined.

One of the conceptual principles for the entity definitions is to define them such that they are scale independent. Since entities exist in the real world at a single scale, it is a reasonable expectation that its definition can be specified to be scale independent. As stored in a database, the digital representation in the deep structure should also be scale independent. However the situation with the spatial objects that are the digital representation of the entity in the database are a different story. Over a short range of scales the graphic surface structure representation may be relatively scale independent, but over wider variations in scale generalization of the representation will usually be required. Such generalizations will usually involve both topological and geometrical principles.

The US standard is unique in that the entity definitions are nonhierarchical by design, although classes can optionally be supported. This conceptual advance distinguishes it from other entity definition structures. This strategy will considerably reduce judgmental error in the initial coding of the data. It should also enhance the efficiency of query retrieval of specific complex object representations. Associated with this advance are the mechanisms for associating attributes with the entity/object representations. More empirical tests need to be carried out to further measure the retrieval efficiencies with these nonhierarchical entity definition structures. Associated with this is a need to develop authoritative numerical tests to assess the quality of a spatial query retrieval for complex entities/objects that also may have similarly involved combinations of attributes. Although this research need sounds somewhat obscure, it is very much a necessary one.

THE EXCHANGE MECHANISM

In SDTS the exchange mechanism is made up of two major parts: the metafile superstructure, and the implementation mechanism. The metafile superstructure comprises a number of exchange modules that contain the codes for spatial data from the three above areas: spatial objects, quality information, and feature attribute information. These modules are designed to be modular in nature so that the source data model information can be decomposed into the exchange modules and associated metadata, transferred via the exchange mechanism, and reconstructed in the data model of the target system. The transfer is accomplished by the exchange mechanism, which in this case is the ISO 8211 'Specification of a Data Descriptive File for Information Interchange' (ISO, 1985). Some of this process has been described by Nyerges (1985) and to a certain extent in Nyerges (1987) who also discusses some research needs associated with this process.

One rather interesting policy question is that of the simplicity of the exchange mechanism versus its complexity versus its flexibility. These are characteristics of every exchange mechanism that transfers spatial data. The reader may wonder why this is a question for policy research rather than normal scientific research. The primary answer is that, at least in the case of SDTS, the scientific design of the SDTS exchange mechanism is robust and flexible enough to accommodate a wide variety of levels of exchange modules. Hence a particular exchange module, say for coordinate systems, could specify 25 coordinate systems that are valid for the transfer, or it could specify a number as small as three and only accept latitude, longitude, Universal Transverse Mercator, and State Plane Coordinate System coordinates. In the latter case, the exchange module is much simpler and so are the source and target metadata converters. But in this case the transfer mechanism would be much less flexible. On the other hand, if 25 coordinate systems were declared directly transferable, the flexibility of the transfer mechanism would increase markedly, but the complexity of the source and target metafile converters would increase tremendously. Now one can see the policy tradeoff between the complexity, simplicity and flexibility. This is not a simple question to answer for a large and diverse community such as the spatial data handling community in the USA, but that kind of question must be answered in order to finalize the standard. The SDTS Technical Review Board has spent months testing, analysing and receiving comments on this matter. Research on the general policy question would be helpful to the overall effort.

In this light more work is needed to ascertain better measures to evaluate the efficiency of the transfer process. Since spatial data are so rich in potential complexity this question is not so easily answered. The first part of the question involves evaluation of the efficiency of the syntax or structure of the database. Although the question of spatial data structures seems complicated, this is actually the easier half of the question. The second half of the question involves the efficiency of the transfer of the semantics or meaning of the spatial data being transferred. Clearly the syntax is fundamental to the process and the semantics are dependent on it to a large extent. However, the semantics are dependent on the consistency of the definitions of the entity and attribute data as they exist in the source and target systems. In other words, the efficiency of the semantics of the transfer are consistent data definitions between the systems, which is independent of the semantics or structure of the spatial data. The general question of syntax/semantic transfer efficiency is an important topic which requires much more research. Very few researchers in the spatial sciences seem to be examining this important topic.

Even though the ISO 8211 database implementation mechanism has been chosen for use with SDTS, more research on the general question of base implementation mechanism is in order. ISO 8211 is a robust, reliable and stable exchange implementation for SDTS. However, spatial data are known to have a considerable richness and potential complexity. More research would help to further evaluate the use of the ISO 8211 implementation mechanism and to see whether any further polishing of the ISO 8211 mechanism itself could perhaps make it even more robust and efficient than it already is. This is a somewhat daunting task in applied research, but this work could verify that ISO 8211 is at peak efficiency now, or perhaps some additional ways may be seen to polish it even further.

Finally, an effort must be made to define, establish and maintain a set of national standardized spatial test data sets that reflect and typify the rich diversity of spatial data in the USA. Such test data sets could be used as a benchmark for the performance of the individual converter modules that decompose into or reconstruct from the metadata transfer module, as well as the efficiency of the metadata transfer process itself. The ability to provide a benchmark for the performance of various converter modules should provide a stable base to compare the speed, accuracy and efficiency of various commercial implementations of converter modules.

INTERNATIONAL COMPARISONS OF EXCHANGE MECHANISMS

As a number of international transfer standards are developed in various parts of the world, a need arises to compare them in terms of conceptual capabilities, comprehensiveness, robustness, speed, accuracy and efficiency, among others. Identifying these capabilities and compatibilities will greatly assist those countries and organizations which are not developing a transfer standard themselves to make intelligent choices between the national standards available and how they might be adapted for the use of that country or organization.

The first step in that process is to identify the conceptual commonalities and compatibilities of the various national standards to transfer various kinds of data structures and data models between various kinds of systems and hardware architectures. This sort of paper evaluation will provide the first set of indications as to what kinds of databases a particular standard may or may not be capable of transferring. The second step will be to determine a benchmark for the capabilities and performance of these national standards in their ability to transfer various kinds of actual spatial data. Once that kind of capability is established, the performance, robustness and efficiency of the transfer standard can be evaluated. This is a practical research need to carry out evaluations for all national transfer standards being developed.

As these various national database exchange standards are developed, there arises a need to identify possible interlinkages between them. The question is whether such national transfer standards could possibly interlink together to transfer certain kinds of spatial databases between them, and hence between various national settings. This question is rather challenging in that this sort of research would require in-depth knowledge of two or more of the transfer standards. After identifying the conceptual commonalities above, the possibilities of establishing operational linkages between the transfer standards would have to be defined. This could be a fairly complicated identification and evaluation process. The feasibility and practicality of such interlinkages would also have to be evaluated.

A different international question is to compare and evaluate the various base implementations being used in the various national standards. SDTS uses the ISO 8211 transfer implementation, while other standards use EDIFACT or other ISO standards. There has been some discussion concerning the robustness, accuracy, speed and efficiency of these transfer vehicles. There is a distinct need to compare these vehicles as to their capabilities and limitations relative to the above criteria. The international spatial data community will benefit greatly from this applied research work.

PRACTITIONER'S SUMMARY

During the past twenty-five years much basic conceptual research work has been carried out that facilitates the development of database transfer standards. Much of this research has been done in analytical cartography and geographic information systems, with varying amounts being carried over from database technology and mathematics. Work on the basic concepts is in a reasonable state now that the notions of real and virtual maps, and transformations between them are more clearly understood. The understanding of cartographic data levels and spatial data models has greatly improved during the period. More recently, the notions of object-oriented structures and hypermaps have been recognized and examined. The work on spatial objects has been advanced enough such that it has been feasible to specify more sophisticated data transfer mechanisms, although many gaps still exist. The idea of pictorial and image databases have been thought of and are being considered. Many operators for the manipulation of spatial objects have been developed in recent decades, mostly for the parametric data domain in 1- and 2-dimensions. The concept of topology has been integrated into the design and operation of spatial data structures with a reasonable development of the theoretical base that supports this work. Initial work has been done on spatial query languages that whets the intellectual appetite for more research in depth. Concepts that relate to data quality have seen much recent development with the specification of SDTS. The ideas in spatial entities have been completely re-examined and thought through once again. The result has been a nonhierarchical definitional structure based on entities and attributes that is unique in the cartographic world. Currently, several interesting data transfer implementation vehicles exist in the form of ISO 8211, EDIFACT, and others. These vehicles have made the development of more comprehensively exchange standards feasible. Finally, the notion of international comparisons of national exchange mechanisms has been thought of and work is just beginning in this area.

PROJECTIVE SUMMARY

The following is a specific listing of research topics that are recognized in this work. As can be seen by the reader, much research work has been done in the past, but there exists the opportunity for much more research of this kind in the future. Analytical cartography is still a rather young branch of the field and one can expect more solid results to be produced if the work is focused on the more pressing topics for research. This list presents these topics in a systematic form. They are:

(1) Extend the transformational work on real and virtual maps.
(2) Extend the understanding and use of Nyerges data levels.
(3) The search for more comprehensive hybrid data structures must be continued.
(4) Object-oriented database concepts are in their infancy in cartography and must be explored in depth.
(5) The richness of hypermaps must be further examined.
(6) The question of the possibility of a unified theory of spatial data must be considered and examined.
(7) 3-D spatial object primitives *must* be developed more completely.
(8) The understanding of 3-D topology *must* be expanded and formal theory that is directly applicable to cartography must be developed, particularly in the irregular solids domain.
(9) Further work on the topological properties of pixels and grid cells is needed.
(10) The notion of topology only objects must be further explored.
(11) Hyperlink relationships need more work.
(12) Pictorial and image objects require further development.
(13) The notion of temporal aspects of spatial data structures is still in its infancy and presents many research opportunities.
(14) The idea of manipulating complex objects needs more work.
(15) True 3-D spatial operators need to be developed.
(16) The idea of data structure independent operators is still immature and requires more work.
(17) The idea of topology only operators should be investigated.
(18) Nonparametric multidimensional spaces should be considered.
(19) True spatial query languages for the retrieval and manipulation of spatial objects is urgently needed.
(20) Better Quality Control methods should be adopted and adapted for spatial database building.
(21) More concise approaches to the reporting of quantitative data quality levels need development.
(22) Better lineage metadata coding schemes should be developed.

(23) The notion of scale independent entity definitions should be extended.
(24) The work on nonhierarchical structures for entity definitions needs expansion.
(25) Rigorous comparisons and performance tests of the various transfer implementation methods such as ISO 8211, EDIFACT, etc, should be carried out.
(26) Better measures for the efficiency of the database transfer process need to be developed.
(27) More work on the concept of simplicity versus complexity tradeoffs for transfer standards is necessary
(28) Syntax/semantic transfer concepts need fuller development.
(29) Standard test data sets that could be used as benchmarks for a transfer process are needed.
(30) International comparisons between various national cartographic database transfer mechanisms is required.
(31) Interlinkages between national transfer mechanisms need to be identified and specified.

As can be seen from the above discussion there are many opportunities for research that relate to the development of spatial database exchange standards. Because these exchange mechanisms and the data they actually transfer touch such a wide base of concepts and theory, the research opportunities offered are really rather diverse and varied. The need for research in this area is more important than many people would assume. The importance of this work cannot be understated because the exchange of spatial data provides a critical element in modern cartography and spatial systems. The major research needs are presented in this discussion. The challenge to the field is to do something about them.

REFERENCES

Abel D J, Wilson M A (1990) A systems approach to integration of raster and vector data and operations *Proc. of the 4th International Symposium on Spatial Data Handling*, 1990 Zürich, IGU Commission on Geographic Information Systems, **2**: 559–66

Ballou D P, Tayi G K (1989) Methodology for allocating resources for data quality enhancement *Communications of the ACM* **32** (3): 320–9

Bouille F (1978) Structuring cartographic data and spatial processes with the hypergraph-based data structure. In: G Dutton (ed.) *Data Structures: Surficial and Multi-dimensional*, Vol. 5, Harvard papers on Geographic Information Systems. Reading, MA, Addison-Wesley

Bouille, F (1984) Knowledge structuring with the HBDS in a cartographic structured expert system. *Technical Papers of the Austra-Carto One Seminar*. Perth, Australia, 12th ICA Conference: 19–26

Chrisman N R (1982) Methods of spatial analysis based on error in categorical maps. Unpublished PhD dissertation, Dept. of Geography, University of Bristol.

Chrisman N R (1989) Error in categorical maps: Testing versus simulation *Proc. Ninth International Symposium on Computer-Assisted Cartography* (*AUTO-CARTO* 9). Falls Church, VA, ACSM and ASPRS: 521–9

Claire R, Guptill S (1982) Spatial operators for selected data structures. *International Symposium on Computer-Assisted Cartography* (*AUTO-CARTO V*). Falls Church, VA, ASP & ACSM: 189–200

Corbett J (1979) *Topological Principles of Cartography*, US Bureau of the Census, Technical Paper No. 48. Washington, DC, US Department of Commerce, Bureau of the Census

Danforth S, Tomlinson C (1988) Type theories and object-oriented programming. *ACM Computing Surveys* **20** (1) : 29–72

Digital Cartographic Data Standards Task Force (1988) The proposed standard for digital cartographic data. American Cartographer **15** (1) 140pp. + 2 appendices

Egenhofer M J, Frank A U (1989) Object-oriented modeling in GIS: Inheritance and propagation *Proc. Ninth International Symposium on Computer-Assisted Cartography* (*AUTO-CARTO* 9). Falls Church, VA, ACSM and ASPRS: 588–98

Frank A U (ed.) (1990) *University of Maine NCGIA* — 18 *Month Report*. Orono, University of Maine, 56 pp

Grosky W I, Mehrotra R (ed.) (1989) Image database management. Dedicated issue of *IEEE Computer* **22** (12)

International Standards Organization (1985) *Specification of a Data Descriptive File for Information Interchange — ISO* New York, American National Standards Institute. Also known as Federal Information Processing Standard — FIPS 123. Gaithersburg, MD, US National Institute for Standards and Technology

Kemp Z (1990) An object-oriented model for spatial data. *Proc. of the* 4*th International Symposium on Spatial Data Handling*, 1990. Zürich, IGU Commission on Geographic Information Systems **2** : 659–68

Laurini R, Millert-Raffort F (1989) Principles of geomatic hypermaps. Paper presented to the 1989 meeting of the Spatially Oriented Referencing Systems Association

Moellering H (1980) Strategies for real-time cartography. *Cartographic Journal* **17** (1): 12–15

Moellering H (1982) The challenge of developing a set of national digital cartographic data standards for the United States. *Proc. of the Spring ACSM Meetings, Denver* : 210–25. Also published in *Papers from the Joint ACSM/ASP Session on Digital Cartographic Data Standards at the* 1982 *Meeting of the American Congress on Surveying and Mapping, Denver, Colorado*. Issues in Digital Cartographic Data Standards, Report 1. Columbus, OH, National Committee for Digital Cartographic Data Standards: 1–15

Moellering H (1984) Real maps, virtual maps and interactive cartography. In: G Gaile and C Wilmott (ed.) *Spatial Statistics and Models*. Dordrecht, Reidel Publishing Co: 109–31

Moellering H (ed.) (1987*a*) *A Draft Proposed Standard for Digital Cartographic Data*, Report 8. Issues in Digital Cartographic Data Standards. Columbus, Ohio, National Committee for Digital Cartographic Data Standards, 316 pp. Now only available as USGS Open File Report 87–308, Federal Center, Building 41, P.O. Box 25425, Denver, CO 80225

Moellering H (1987*b*) GIS research needs identified during a cartographic standards process: Cartographic object definition and manipulation. *Proc. International GIS Symposium*, 1987, Washington, DC, Association of American Geographers: I – 331–I – 337

Moellering H (ed.) (1991) *Spatial Database Exchange Standards: Current International Status*. International Cartographic Association (in press)

Nyerges T (1980) Modelling the structure of cartographic Information for Query Processing. Unpublished PhD dissertation, Dept. of Geography, Ohio State University

Nyerges T (ed.) (1984) A progress report: Working Group on Data Organization. In: Moellering H (ed.) *Digital Cartographic Data Standards: Examining the Alternatives*. Report 4, Issues in Digital Cartographic Data Standards. Columbus, Ohio: National Committee for Digital Cartographic Data Standards: 73–92. Now only available as USGS Open File Report 87–304, Federal Center, Building 41, P.O. Box 25425, Denver, CO 80225

Nyerges T (1985) An interim proposed standard for digital cartographic data organization: Supporting documentation. In: Moellering H (ed.) *Digital Cartographic Data Standards: An Interim Proposed Standard*. Report 6, Issues in Digital Cartographic Data Standards. Columbus, Ohio, National Committee for Digital Cartographic Data Standards: 67–111. Now only available as USGS Open File Report 87–306, Federal Center, Building 41, P.O. Box 25425, Denver, CO 80225

Nyerges T (1987) GIS research needs identified during a cartographic standards process: Spatial data exchange. *Proc. International GIS Symposium*, 1987, Crystal City, Virginia: I – 319–I – 330

Pascoe R T, Penny J P (1990) Construction of interfaces for the exchange of geographical data. *International Journal for Geographical Information Systems* **4**(2): 147–56

Peuquet D (1983) A hybrid structure for the storage and manipulation of very large data sets. *Computer Vision, Graphics and Image Processing* **24** (1): 14–27

Peuquet D (1984) A conceptual framework and comparison of spatial data models *Cartographica* **21** (4): 66–113

Ross E (1987) A comparative analysis of the vaster data structure. Unpublished MA paper. Dept. of Geography, Ohio State University, 77 pp. + appx.

Rugg R, Schmidt W (1987) GIS research needs identified during the cartographic standards process: Features. *Proc. International GIS Symposium*, 1987, Washington, DC, Association of American Geographers: I – 307–I – 318

Spatial Data Transfer Standard Technical Review Board (1990) *Spatial Data Transfer Standard*, July, 1990 draft, three parts, 280 pp

TeSelle G (ed.) (1986) *Federal Geographic Exchange Format*, Version 1.08, Final Draft. Reston, VA, Federal Interagency Coordinating Committee on Digital Cartography, 52 pp.

Tobler W (1979) A transformational view of cartography. *American Cartographer*, **6** (2): 101–6

Tobler W (1984) Application of image processing techniques to map processing. *Proc. International Symposium on Spatial Data Handling*, Vol. I. Zurich, Switzerland, Symposium Secretariat, Dept. of Geography, University of Zurich-Irchel: 140–5

White M (1979) A survey of the mathematics of maps. *Proc. International Symposium on Computer-Assisted Cartography* (*AUTO-CARTO* 7). Falls Church, VA, ASP & ACSM: 82–96

White M, Griffin P (1979) Mapping without coordinates. In: R. Schmidt (ed.) *Proc. URISA*, 1979. Chicago, IL, URISA: 431–42

Chapter 8

VEHICLE NAVIGATION SYSTEMS

Hinrich Claussen
Robert Bosch GmbH Dept. MC/ENS1, Hildesheim, FRG
and
David M. Mark
National Center for Geographic Information and Analysis, Department of Geography, State University of New York at Buffalo, Buffalo, New York 14261, USA

INTRODUCTION

Vehicle navigation may be defined as the guidance of a vehicle from a start point to a destination. Position-keeping is an important aspect of navigation, as is route planning; all of these aspects will be considered in this chapter. The methods applied in modern navigation have been available since people began to sail the oceans, but the equipment has been improved over the centuries. Recently, developments in computer and communication technology have had a great influence on navigational methods. This is particularly so for car navigation. Because positioning and route finding for vehicle navigation must be performed in real time with equipment which is compact in size, the decreasing size and costs and increasing speed of computers have only recently made the development of computer-based on-board *Vehicle Navigation Systems* (VNSs) practical. The developments mentioned above coincide with a large demand for these systems, which exists because of an increasing volume of traffic. A major goal in these circumstances is to enhance the management and the safety of road traffic.

Maps have long been the most important navigational aid for both shipping, aviation, and driving, functioning as both a database and as an aid to visualization. Automated vehicle navigation systems have come into existence primarily due to the progress in digital cartography, and thus VNSs represent an important area for advanced research in cartography, even if some systems do not include graphic map output. The importance of VNSs is further indicated by the fact that international conferences on this topic were held in 1989. First, the NATO Advanced

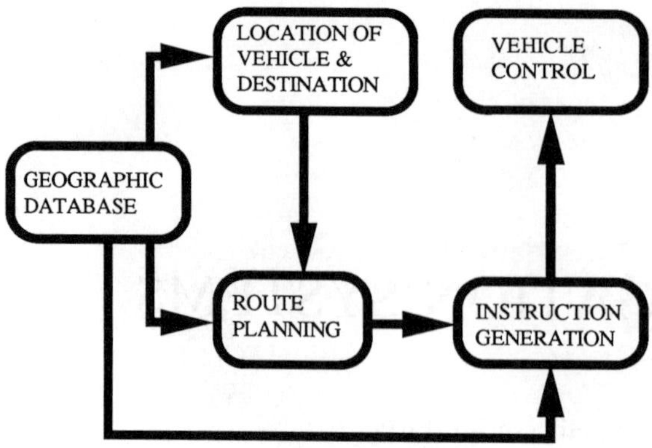

FIGURE 1 Conceptual model of the components of a Vehicle Navigation-Aid System and the information flows between them (Source: Mark 1989)

Research Workshop was held in Denmark in August (Pau, 1990); then, in September 1989, the IEEE Vehicular Technology Section organized such a meeting in Toronto (IEEE 1989), and plans another such meeting for 1991.

Recently, Mark (1989) presented a conceptual model for Vehicle Navigation Systems. In order for vehicles to travel purposefully through geographic space, five functions must be performed:

1. *Geographic Database.* Information about the configuration of geographic space, and about paths and objects within it, must be available.
2. *Location of Vehicle and Destination.* The current location must be known within the reference frame provided by the database (1); furthermore, the location of the destination must also be known.
3. *Route Planning.* A route from the current location to either the exact destination, or to the neighbourhood of the destination, must be determined or recalled.
4. *Instruction Generation.* The action to be taken at the next decision point along the route, together with some means to recognize that decision point, must be available.
5. *Vehicle Control.* The vehicle's travel through the space must be controlled.

The interrelations among these five functions are illustrated in Figure 1. The task for the designer of a VNS can then be defined as assigning the various functions to computer systems, other electronic devices, and human minds, developing methods by which the components may communicate, and designing or training those components. Furthermore, the minds or the machines could be either within the vehicle, or elsewhere.

Mark (1989) gave a number of examples of how the functions could be divided among the human and other components of the system. For example, when a driver, alone in a car, uses a printed road map to navigate in unfamiliar territory, the map acts as the geographic database, and the driver's mind performs all other functions using input from the map and from the environment. On the other hand, when a driver uses written directions compiled by another person, the driver merely follows the instructions (with some location-monitoring) and controls the vehicle; the preparer of the directions had already used a geographic database (personal knowledge, or a map, or some other source), had performed route planning, and had generated the instruction.

In a fully-autonomous vehicle, all of the functions identified in Figure 1 would be performed by computers and other devices on board the vehicle. However, in this paper, we will not discuss such vehicles, and instead will consider only vehicles with human drivers, with the vehicle control function performed by that driver. In the remainder of this paper, then, we will discuss how the other four functions can be performed, and will review relevant issues concerning communication among the components.

GEOGRAPHIC DATABASE

Geographic database requirements for VNSs depend a great deal on the type of system. Key issues are: required data content (data models) to support vehicle navigation; data structures to support sufficiently rapid database access; and data sources.

Digital Map Requirements

All functions of a navigation system make their own particular demands on the on-board database, which may be termed the *digital map*. The map-matching function used in some systems requires geometric information about the road network with sufficient accuracy; this information is also needed for calibration of vehicle parameters.

If the system is to provide a route planning function, the topology of the road network must be represented in the database. This includes information such as whether roads are directly connected at a particular intersection or if they cross each other by means of a bridge or viaduct, etc. The performance of the route planning function can be enhanced with the help of a road classification; that leads to a network or valuated graph, a graph in which each edge has an associated weight. Travel time can then be a criterion for the optimal route. If regions are just to be passed through then roads of less priority can be ignored in order to save computation time. In this case the detailed structure of intersections is not important, but these constructions can be aggregated to one composite

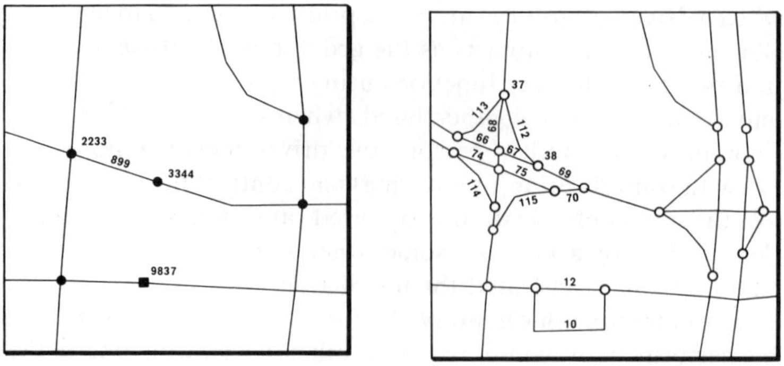

FIGURE 2 Simple and aggregated objects

object as shown on Figure 2. Also, knowledge about traffic restrictions and prohibitions, such as one way roads, prohibited turns, etc, should be taken into account to avoid planning and suggesting illegal routes. Dimensional restrictions are particularly useful for truck drivers (ie height, width, length).

For clear driving instructions the route guiding function needs an aggregated representation of the road network. Road sections form roads; internal links and junctions form intersections, etc. Special features like roundabouts (traffic circles) require special instructions. For this reason, the route guider has to know the road elements that belong to a roundabout. Furthermore, the information represented on signposts is useful, because guiding instructions may then refer to it.

A further important function of a navigation system that must use the database is address conversion. In most countries the common way is to specify addresses in terms of town names, street names, and house numbers. But the database system however uses internal addresses specified, for example, by identification-numbers of roads or other objects. Uniqueness (or non-uniqueness) of addresses also plays a significant role. To guarantee uniqueness of a street name outside the context of a particular municipality, municipality names and also higher order administrative areas are required. The same holds for objects like restaurants, cinemas, petrol (gasoline) stations, etc, assuming that they are recorded. The knowledge of house numbers enables the specification of a more precise location on a street.

Finally, the database should also contain a number of land use units, railways, rivers, canals and other landmarks. Such landmarks can be plotted on a graphic map display, to improve the readability of map images produced in VNSs that have a position display function (McGranaghan, 1987). The landmarks would also play a key role in the production of clear verbal driving directions (see Mark and McGranaghan, 1986).

Data Model and Data Content

Locations are given by pairs of coordinates (points). They can be attached to nodes and to chains (lines represented by ordered sets of points). In the case of chains they describe the shape by means of intermediate points. Geometry can be attached to features. Geometry of point features such as landmarks can be attached directly to the representations of those features.

One of the most important factors for many applications and systems is not the geometry (coordinate information) but how the data are topologically structured. Topology enables the description of connections between neighbouring features. Finding a path through a network is a topological calculation. For map-matching purposes, the topological property of connectedness is required. Topologically, the data are stored as nodes (0-cells) and chains (1-cells). Connections are indicated by start and end nodes of chains or by a pointer from nodes to connected chains. Nodes are located at all intersections of chains. On the object level, connections of road sections are explicitly indicated.

Attributes are essential in a road database. Important attributes for roads are: road class, one-way flow and turn restriction, and meanings of traffic signs. Street names and highway numbers also are essential for some VNSs. Administrative areas may have area codes and type names. For buildings and services, attributes like brand names and opening hours may increase serviceability. Some of them can be captured directly from annotations and symbols on paper maps. But others, such as content of traffic signs and house numbers, require field work. After the survey, these attributes must be attached to the data gathered from maps. This is performed by pointers from features to attribute identifications.

Apart from relationships between different data types, such as attributes and features, relationships between different feature types can be indicated explicitly. These relationships, where more than one feature is involved, concern prohibited turns, right of way and signpost information. Other relationships are created to give the possibility of describing situations in which the logical tie between features is not equivalent to their spatial one, eg a building has its entrance on another road than the one which is the nearest. A further type of relationship is created to describe flyovers (overpasses); if the roads do not connect, then which road is the upper and which the lower one.

In addition to spatial information, a data set has to contain global information that is needed for correct interpretation of data sets. Such 'meta'-data also can keep external documentation to a minimum. Each data set can be subdivided into sections and layers, whereby a section is a spatial subdivision and a layer a logical subdivision. The data set itself and the subsets must be preceded by a corresponding header, which identifies and briefly documents the data collection and manipulation methods. Also important is the group of bibliographic data which explain the lineage by means of identification of source documents.

A directory describes the content, size, and division into subsets of a data set. Geometric data must be provided according to a unique reference system, but it is essential that geodetic data are indicated. These must provide information about items such as the horizontal datum, projection system, projection parameters, origin of the coordinate system, coordinate off-sets, vertical datum, geoid undulation, and the magnetic declination. The quality of a data set should be given in terms of geometrical accuracy, currency ('up-to-dateness'), completeness of features and attributes, and error rates in attribute values.

According to the conceptual data model above, an exchange format has been created within the DEMETER (Digital Electronic Mapping of European TERritory) project. The goal of the exchange format is to meet the following requirements (Claussen et al. 1989):

- —It should have enough representational power to cover the special needs of a variety of navigation systems.
- —It should have the ability to represent topologically structured data.
- —It should allow independence between the representation of the semantic, topological, and geometric information of a particular feature.
- —It should be open-ended so that it could be extended with new fields, records, feature codes and attribute types, without the need to redesign the format completely.

Data Sources

The main data sources of digital map data for car navigation systems are either existing databases created for other purposes, or analogue road or topographic maps. Acquisition of road data for vehicle navigation systems from maps, with the required topological, attribute, and name data, is very costly and time consuming. Both time and costs can be decreased if this work is done only once, in a way which allows the data to be accessed by several systems. This, however, requires standardized methods for acquisition and exchange of data. In the USA, large public-domain databases exist at the Bureau of Census and at the US Geological Survey. The DIME files of the pre-1990 US Bureau of the Census have been adapted to vehicle navigation by companies such as Geographic Data Technologies Inc. (see Cooke 1985). On the other hand, ETAK Inc. has obtained most of the data used by their system by scanning the printed, public-domain topographic maps of the US Geological Survey (White 1985, 1987).

In Europe, the DEMETER project (mentioned above) has been

established, and addresses data needs as well as formats and data models. Its goal is to create standards for digital map data which will be primarily used for car navigation purposes. The first result of the work in this project is a standard named GDF 1.0 (Geographic Data File Release 1.0). The standard consists of a Specification of Data Content (SDC), a Specification for Data Acquisition (SDA), and an Exchange Format (EF). The data model of GDF is feature oriented. A feature is a formalized entity made to represent a topographical object.

A group of features that are strongly related is called a layer. The most important layer for car navigation is the road layer. Other layers include: administrative areas, buildings, railways, and waterways. Features have both quantitative and qualitative properties; these are described by means of attributes. Interrelationships can be recorded explicitly. Shape and location of features are expressed by means of geometric data.

LOCATION OF VEHICLE AND DESTINATION

Several industry leaders are engaged in the development of special electronic components for automobile navigation. These components provide vehicle navigation and step by step route guidance, aiming to create a safer and more relaxed driving environment.

The minimum requirement for any system (driver plus instruments) is the knowledge of the position of the car at any time and the destination in the same coordinate system, in order to enable the computation of direction and line-of-sight distance to the destination. If this is not known, the system, or the driver, is 'lost'.

Currently, a wide range of systems with a variety of performance capabilities is emerging. These systems either are under development or are already available on the market. They can be divided into two broad types; map-matching and non map-matching systems. Non map-matching systems work with the support of analogue map data only, whereas map-matching systems use digital map data in the car as support for positioning and route guidance.

The method of positioning leads to another classification. Positioning can be performed either by support of infrastructure or autonomously, as shown on Figure 3. For autonomous systems the equipment needed for determination of the actual location is installed in the car.

Infrastructural-supported systems are based on support from a satellite system, proximity beacons, or radio communications. Inputs from one or more of these sources are necessary for the determination of the automobile's location. Proximity beacons can also be used for transmission of actual traffic data, such as traffic jams, and road works

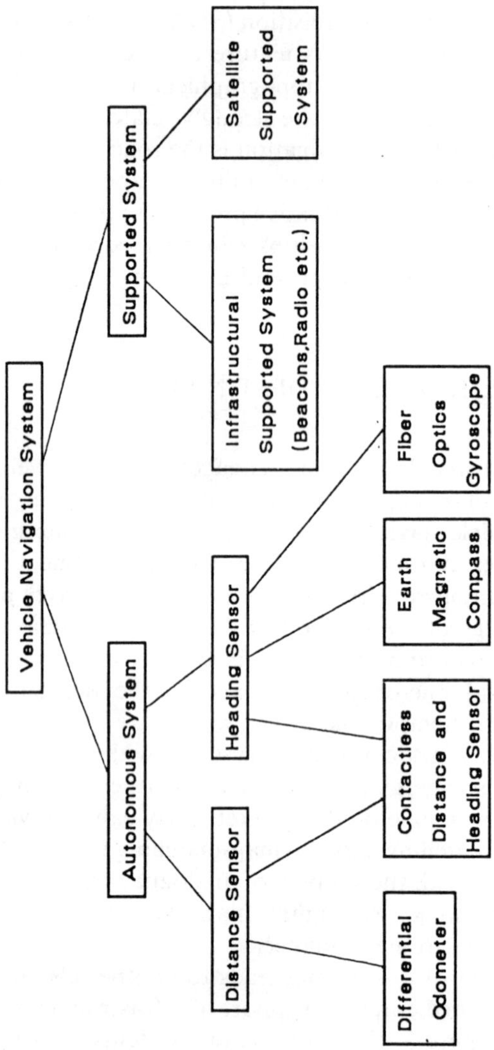

FIGURE 3 Vehicle Navigation Systems

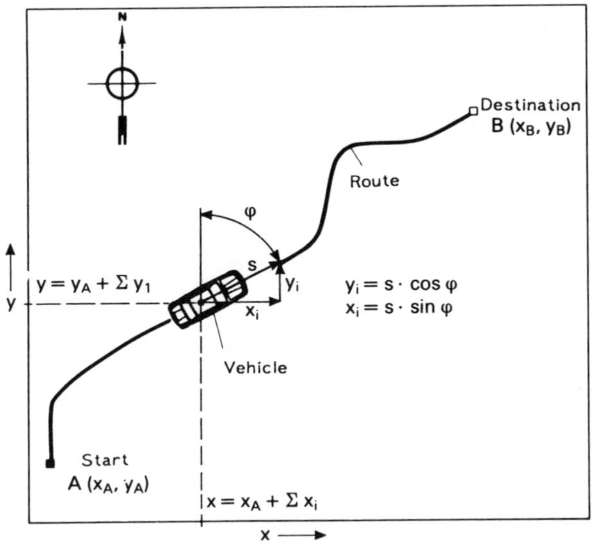

FIGURE 4 Dead reckoning (Source: Heintz, Knoll 1986)

(construction areas). But the advantages of having the most recent information about traffic events is offset by the considerable effort and cost needed to provide infrastructure and by the disadvantage that the system has much less independence.

Autonomous systems adopt the dead reckoning method, well known from ship navigation and aeronautics. This is used either on its own or in conjunction with map-matching procedures. The principle is that vectors obtained by increments of distances and directions are added continuously and provide the actual position, as shown in Figure 4. Dead reckoning requires a distance sensor and a heading sensor. Several distance sensors have been developed. As a source of signal for distance increments, pulse generators mounted on the non-driven wheels of a car can be used. They deliver a pre-defined number of pulses per rotation for the left and the right wheel. Usually these sensors are used for cars with an anti-lock braking system. For positioning purposes they also must manifest static behaviour, because evaluable pulses are needed down to very low speeds (1 km/h) . If the sensors are paired, recognition of the sense of rotation is possible and position errors caused by reverse movement can be avoided. The variation in direction can be derived from the difference between the distances covered by the inner and the outer wheel in a curve, as these values are proportional to each other as is shown on Figure 5. Errors may appear especially if the radius of the wheels or the wheel track is not well known.

Earth magnetic field sensors (flux-gate magnetic compasses) measure an absolute heading to a natural fixed point (magnetic pole). However the

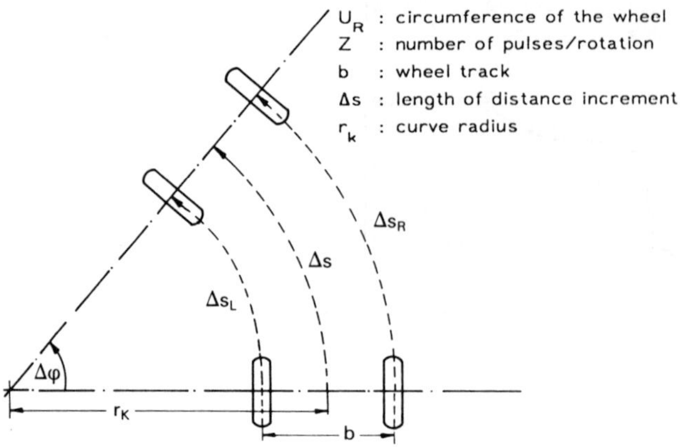

$$(1) \quad \Delta s = r_K \cdot \Delta\varphi$$

$$(2) \quad \Delta s_R = \frac{U_R}{Z} \cdot Z_R = (r_K + b/2) \cdot \Delta\varphi$$

$$(3) \quad \Delta s_L = \frac{U_R}{Z} \cdot Z_L = (r_K - b/2) \cdot \Delta\varphi$$

$$(2)-(3) \quad \frac{U_R}{Z}(z_R - z_L) = b \cdot \Delta\varphi \rightarrow \Delta\varphi = \frac{U_R}{b \cdot Z} \cdot \Delta z$$

$$(2)+(3) \quad \frac{U_R}{Z}(z_R + z_L) = 2\, r_K \cdot \Delta\varphi \rightarrow \Delta s = \frac{U_R}{Z} \cdot z$$

FIGURE 5 Variation in direction (Source: Neukirchner, Zechnall 1986)

measurement with a flux-gate magnetic compass is prone to error caused by local influences, such as the starting of electric trams, or the magnetic variation caused by power lines and steel bridges. Wheel sensors are insensitive to the short term distortions which affect magnetic compasses. As a compromise, filtering and addition of the interfering signals of both sensors provides improved direction values and increases the reliability of the car's location.

Non Map-Matching-Systems

Two European examples of non-map-matching systems are CARPILOT (Bosch-Blaupunkt) and CITYPILOT (VDO). These systems use an Earth magnetic field sensor and a differential odometer. A North American example is CLASS (Chrysler). Examples of other systems are given by French (1987). In these systems, the driver has to enter the origin and destination coordinates. In the system developed by the company VDO, a light pen can be used to read the bar-coded location of the start and the destination on special maps. The on-board computer in turn calculates directions and line-of-site distances to the driver's destination. On a simple liquid crystal display, the computer presents the results of its calculation process. Utilizing either of these two systems will allow the driver to reach a specified destination within an error range of about 2 to 3% of the distance travelled.

Map-Matching Systems

The positioning method described above is still affected by both systematic and random errors. If the sensors are properly calibrated, then most of the systematic errors can be avoided. The remaining errors can be regarded as random errors, and their size can be expected to grow with the length of the journey. If a digital map database is available on a mass storage medium in the car, position correction can be made. By comparing a number of fixes obtained by dead reckoning with the topology and geometry of the map data, small but potentially cumulative errors in the dead reckoning can be eliminated, and the actual position can be corrected.

ROUTE PLANNING

Many systems do not provide advice on routes. The current vehicle location is displayed on a graphic map, and the driver or navigator must interpret the graphic and determine the route. VNSs can be divided into those which perform route guidance and those which do not.

Systems without Route Planning and Route Guidance

Systems without a Route Planning Component require:

a) input of the preferred route;
b) output of route information.

For the input of the preferred route, a small digitizer can be used, with which the selected route on a conventional paper map is digitized. The navigation computer represents the route graphically on a display. A symbol marks the location of the automobile. The locations are determined by dead reckoning and map matching as described above. A more advanced system of this class is Navigator, offered by ETAK on the American market (Zavoli et al. 1985, Kelley 1986). The system continuously displays the location of the automobile and the surrounding road network on a CRT screen. The digital map data are stored on 3·5 Mbyte cassettes.

On the European market, a product called TRAVELPILOT is available with the same performance, but with a CD-ROM with a capacity of 540 Mbytes as the storage medium. The destination can be entered in the form of towns and street names, by a menu displayed on the CRT. The locations of the origin and destination are then illustrated on the screen. As with the ETAK Navigator, the display rotates corresponding to the heading of the automobile, so that the road network is always observed from the driver's view. Different map scales can be shown according to the distance to be travelled and the detail required to be represented.

Systems with Route Planning and Route Guidance

A system of this class is required to answer the following questions for navigation:

1. What is the actual location of the automobile?
2. What is the direction to driver's destination.
3. What direction or road must the driver choose at the next decision point to reach his destination by an optimal route?

and perhaps a fourth:

4. What is the line-of-sight distance to driver's destination?

To recommend an optimal direction to a destination, route planning is required. Algorithms for this purpose are available in graph theory. Vehicle navigation systems have special requirements, because routes must be planned in real time. If the driver does not follow the recommendation, the route planner has to offer a new optimal route. The algorithm of Ford and Moore (Domeschke 1972) provides routes in valuated graphs. The valuation stems from attaching the needed travel time to road sections. (If travel times are not available, link lengths can be substituted.) The planning of the route is performed in a hierarchical way. If the route is planned from the destination to all road sections, the result can be a complete table of routes leading to the destination. If the recommended route is not followed, the next one can easily be identified.

Prototypes of systems with this performance are EVA (Elektronischer Verkehrslotse für Autofahrer — Electronic guide for automobile drivers) developed by the German company Bosch and CARIN (Car Information and Navigation System) developed by Philips (Netherlands) (Thoone 1987). For safety reasons, both systems use synthetic voice generators to output driving instructions (see section on 'Instruction Generation', below). With the aid of a communication system, the driver enters the destination of the trip. The actual location of the vehicle is constantly known, even while parked. The location and route planning system provide input to the navigation system. The on-board computer processes the input data, calculates an optimal route, and prepares a step-by-step route guidance. The communication system presents this information to the driver by synthetic voice and by visual display.

Infrastructure-assisted Systems

Whereas this approach is now inactive in the USA (French 1987), further development has been undertaken in Europe and Japan. A demand on these systems exists where traffic situations change very frequently, because they allow traffic dependent vehicle guidance. In West Germany the ALI System (Autofahrer Leit-und Informationssystem — Automobile Driver Guidance and Information System) was proposed in 1970. Subsequently a field test was prepared and was carried out in 1979–82 in an autobahn test network of about 150 km in length. The communication between the car and a central control computer is performed by means of induction loops. Information such as speed, type of car, desired destination, etc, are transmitted to the computer, where the actual traffic situation is derived. Considering this, the individually recommended route is calculated. Driving instructions are then sent to the car and given to the driver in visual form whenever an induction loop is passed at an intersection or highway exit. In addition to that, warnings on road conditions like traffic jam, ice, fog, etc, are transmitted into the car and displayed on a vehicle unit. The field test was successful and proved a first approach towards an individual guidance system.

Further development was evoked by the decreasing costs of semiconductors, as it became possible to have more computer intelligence in the on-board computer. The system ALl-Scout consists of infrastructure and in-car equipment. This system is a result of an evaluation of existing navigation and guidance systems. Infrared beacons are used for transmission of information. They are installed at important intersections of the road network. The actual traffic situation can be derived from travel times sent to a central traffic computer. Traffic dependent driving instructions are then calculated and sent to beacon control units. The instructions are received by the on-board computer in the car and given to the driver visually or acoustically when passing a beacon. The vehicle's position is obtained by dead reckoning and map-matching methods.

The ALl-Scout system is being tested in the large scale field test LISB (Leit- und Informationssystem Berlin — Guidance and Information System Berlin). The test covers the entire street network of West Berlin. Two-hundred and twenty intersections are equipped with infrared beacons, which communicate with 700 test drivers. Auto Guide, a system similar to ALl-Scout, is being developed by the Automobile Association (AA) in Great Britain (Catling and Belcher 1989), and infrastructure-based systems are also under consideration in the USA (Case 1989).

INSTRUCTION GENERATION

One of the most critical functions in a Vehicle Navigation System is the determination of a set of *actions* to be taken, together with means to recognize when and where the actions should be performed. Turns may be the most important actions, but other actions, such as lane changes, often are needed. Kuipers (1978, 1983*a*, 1983*b*) models spatial knowledge of routes as series of what he called view–action pairs, where action was defined as above, and view was used to represent the total of sensory inputs that allow the traveller to recognize the decision point. In fact, Kuipers pointed out that if all views of the environment are unique, then an unordered collection of view–action pairs is sufficient to represent a route, since after an action is performed, the environment will provide the next view. A person with such disconnected spatial knowledge might say: 'I can't tell you how to get there, but I could show you'.

For a driver alone, in a car equipped with a VNS, the instructions can be generated by either the VNS or the mind of the driver. Most existing systems present information to the driver through maps displayed within the car. For these, as for paper road maps, the driver's mind must do the work needed to produce instructions. The task involves map-reading as well as spatial inference, based on mental models (schemata) of the type of environment within which travel is taking place. For example, in a typical midwestern city in the USA, the roads form a rectangular grid; assuming such a grid provides a powerful set of inference heuristics. A North American driver who uses those heuristics in a European city may, however, have considerable difficulty. A display-based system raises many map-design issues, most of which have not yet been addressed by carefully-designed experiments (McGranaghan, 1987).

The alternative is to have the VNS determine the instructions, and then present them to the driver. The presentation could be through natural language, or through signals. Perhaps the most comfortable form of navigation assistance is that provided by a passenger who is a local resident and who verbally directs the driver to the destination. This suggests that if a VNS could mimic that situation sufficiently, this would be an excellent VNS strategy. Experiments performed by Streeter and her colleagues (Streeter et al. 1985) indicated that her subjects could navigate more effectively when using a tape-recorded set of verbal instructions than when using a customized road map. Considerable theoretical and practical work on systems to compute paths and generate verbal directions has been conducted (Elliott and Lesk 1982, Mark 1985, 1987, Davis 1986, Ma 1987, Davis and Schmandt 1989), but further work is needed, both on these problems and on replications of Streeter's experiments in other situations and with more subjects.

One of the many difficulties in the case of verbal directions is that the VNS must select the route. Although route planning might be desired on map-based systems, it is not required; however, the VNS cannot generate instructions without a route that either is entered interactively by the user, is stored in the system, or is determined 'on the fly' by the computerized VNS itself.

As noted above, instructions might also be presented through non-verbal signals on the vehicle dashboard or elsewhere in the driver's field of view. The signals could be arrows or blinking lights to indicate up-coming turns. The signals could change colour, or blink faster, to indicate a turn at the next intersection. This could be accompanied by sounds, possibly including words.

And to further complicate matters, individual differences in preferences and perhaps performance almost certainly exists (Gould 1989*a*, 1989*b*). And, there are notable gender differences in both spoken and written driving directions (Freundschuh et al. 1990), suggesting possible differences in performance when using such vehicle navigation information. Systems of the future will almost certainly have to take such variations into account, and either test new users and configure the system accordingly, or allow users to customize the systems themselves.

PRACTITIONERS' SUMMARY

Cartographers have long know that location maps in public places are most useful if the current location (location of the sign) is indicated by a 'you are here' symbol. Also, such maps are most effective when oriented so that the 'away' direction is up. Unfortunately, these map-design principles cannot be applied to portable maps (such as road maps) printed on paper or other static media. Computerized vehicle navigation systems now allow such maps to be presented dynamically in moving vehicles, with the current heading 'up' and the current position marked. Several technologies have been used to achieve this, and costs are now about US $1000 at the most. However, the same advanced technology that allows this also permits vehicle navigation advice to be provided to drivers in verbal or signal forms. Research is needed to compare the effectiveness of these forms, and to improve the quality of each type of information.

Automated vehicle navigation aid systems have strong needs for digital cartographic data. Automated systems for private cars, delivery fleets, and emergency vehicles require accurate and complete street databases, with efficient data structures for real-time retrieval of such data. This in turn has provided considerable impetus for commercially-produced and improved digital street data, and can be expected to be a driving force in digital cartographic database construction and update for several years to come.

PROJECTIVE SUMMARY

The advanced technology needed for vehicle navigation systems has already been developed, and some very promising tests have been performed. In limited urban areas, infrastructure-supported systems may provide an appropriate method for fleet management. However, for individual guidance purposes, autonomous systems are probably more appropriate. The pre-requisite is that digital road data can be provided for extended areas, and such data must include traffic-relevant attributes which must be captured by field survey. This requires a major effort, and is only likely to occur through cooperation among the various industrial and governmental agencies which are interested in vehicle navigation. The databases must be updated continuously, and in the future, actual traffic data will be required.

In this field, the Radio Data Signal (RDS) is very promising. The possibility of communication from infrastructure to vehicles will be a first step to enhance the benefits of the systems described. However, vehicle-to-infrastructure data communication could provide destination information to central traffic management systems for planning optimal traffic flow. Finally, for wide propagation of car navigation systems standardization is required in the fields of navigation, digital maps and mobile data communication.

There is much research to be done by human factors engineers, geographers, cognitive scientists, and especially cartographers. If systems are to present maps to the user, research is needed to establish ideal design criteria. But research also is needed to determine when navigation assistance should be presented to drivers as maps, when it should be done through synthesized speech, and when it should be presented through signals. No doubt all of these approaches are useful, but the relative advantages, and the influences of individual differences, cognitive style, geographic setting, and vehicle environment remain to be determined.

ACKNOWLEDGEMENTS

Mark's contributions to this paper represent part of Research Initiative 2 ('Languages of Spatial Relations') of the National Center for Geographic Information and Analysis, supported by a grant from the National

Science Foundation (SES–88–10917); support by NSF is gratefully acknowledged. The bibliography of this paper was derived in part from a report by Mark et al. (1989); contributions to that report by all of its co-authors are appreciated.

REFERENCES AND BIBLIOGRAPHY

Case E R (1989) An advanced driver information system concept for North America: A MOBILITY 2000 report. *Proceedings, First Vehicle Navigation & Information Systems Conference (VNIS* '89). IEEE Vehicular Technology Section, Toronto, Ontario, Sept. 11–13, 1989: 489–93

Catling I, Belcher P (1989) Autoguide — route guidance in the United Kingdom. *Proceedings, First Vehicle Navigation & Information Systems Conference (VNIS* '89). IEEE Vehicular Technology Section, Toronto, Ontario, Sept. 11–13, 1989; 463–6

Claussen H (1989) GDF — Ein Austauschformat für geographische Daten. *Nachrichten aus dem Karten und Vermessungswesen.* Frankfurt am Main, Germany, Reihe 1, Heft 103: 37–44

Claussen H, Heres L, Lichtner W, Schlögl D (1988) Digital Electronic Mapping of European Territory (DEMETER). *Proceedings 18th International Symposium on Automotive Technology & Automation.* Florence Italy, 1988

Claussen H, Heres L, Lahaije P, Lichtner W, Siebold J (1989) GDF, A proposed standard for digital road maps in car navigation system. *Proceedings, First Vehicle Navigation & Information Systems Conference (VNIS* '89). IEEE Vehicular Technology Section, Toronto, Ontario, Sept. 11–13, 1989

Cooke D F (1985) Vehicle navigation appliances. *Proceedings, 7th International Symposium on Computer-Assisted Cartography*: 108–15

Davis J R (1986) Giving directions: A voice interface to a direction giving program. *Proceedings*, 1986 *Conference*, American Voice I/O Society, Sept. 1986: 77–84

Davis J R, Schmandt C M (1989) The back seat driver: Real time spoken driving directions. *Proceedings, First Vehicle Navigation & Information Systems Conference (VNIS* '89). IEEE Vehicular Technology Section. New York, The IEEE Society: 146–50

Domeschke W (1972) Kürzeste Wege in Graphen: Algorithmen, Verfahrensvergleiche. Mathematical Systems in Economics, **2**. Meisenheim

Elliott R J, Lesk M E (1982) Route finding in street maps by computers and people. *Proceedings, AAAI–82 Conference*

French R L (1987) Automobile navigation in the past, present, and future. *Proceedings, 8th International Symposium on Computer-Assisted Cartography*: 542–51

Freundschuh S M, Mark D M, Gopal S, Gould M D, Couclelis H (1990) Verbal directions for wayfinding: Implications for navigation and geographic information and analysis systems. *Proceedings, 4th International Symposium on Spatial Data Handling*, July 1990 **1**: 478–87

Gould M D (1989*a*) Considering individual cognitive ability in the provision of usable navigation assistance. *Proceedings, First Vehicle Navigation & Information Systems Conference* (*VNIS* '89). IEEE Vehicular Technology Section, Toronto, Ontario, Sept. 11–13, 1989: 443–7.

Gould M D (1989*b*) Exploring cognitive tests as a means to categorize drivers by navigation ability. *The East Lakes Geographer* **24**: 54–62

Heintz F, Knoll M (1986) Kraftfahrzeuginformation: Ergonomie und Technik. Stuttgart, Germany, Bosch-Technische Berichte, **8** (1/2): 32–46

IEEE (1989). *Proceedings, First Vehicle Navigation & Information Systems Conference* (*VNIS* '89). IEEE Vehicular Technology Section, Toronto, Ontario, Sept. 11–13, 1989

Kelley J (1986) With Stan Honey's high-tech road maps, a motorist never has to say, 'I'm lost'. *People*, 31 March 1986

Kuipers B (1978) Modeling spatial knowledge. *Cognitive Science* **2**: 129–53

Kuipers B (1983*a*) The cognitive map: Could it have been any other way? In: Pick H L Jr. and Acredolo L P O (ed.) *Spatial Orientation*: *Theory*, *Research*, *and Application*. New York, Plenum Press: 345–59

Kuipers B (1983*b*) Modeling human knowledge of routes: Partial knowledge and individual variation. *Proceedings*, *AAAI* 1983 *Conference*, The National Conference on Artificial Intelligence: 1–4

Lahaije P D M, Wester R H H (1988) Efficient road-map management for a car navigation system. *Philips Journal of Research*, **43** (5/6).

Lichtner W (1988) The integration of digital maps in vehicle navigation systems. *Proceedings of the* 13*th International Cartographic Conference*, Morelia, Mexico, Oct. 12–21, 1987 **4**: 1–10

Ma P (1987) An algorithm to generate verbal instructions for vehicle navigation using a geographic database. *The East Lakes Geographer* **22**: 44–60

Mark D M (1985) Finding simple routes: 'Ease of description' as an objective function in automated route selection. *Proceedings*, *Second Symposium on Artificial Intelligence Applications* (IEEE), Miami Beach, Dec. 1985: 577–81

Mark D M (1987) On giving and receiving directions: Cartographic and cognitive issues. *Proceedings*, 8*th International Symposium on Computer-Assisted Cartography*, Baltimore, Maryland: 562–71

Mark D M (1989) A conceptual model for vehicle navigation systems. *Proceedings*, *First Vehicle Navigation & Information Systems Conference* (*VNIS* '89). IEEE Vehicular Technology Section, Toronto, Ontario, Sept. 11–13, 1989: 448–53

Mark D M, McGranaghan M, (1986) Effective provision of navigation assistance for drivers: A cognitive science approach. *Proceedings*, *Auto-Carto London*, London, England, Sept. 14–19, **2**: 399–408

Mark D M, Gould M D, Freundschuh S M, Egenhofer M J, Kuhn W, McGranaghan M, Svorou S (1989) Working Bibliography on 'Languages of Spatial Relations'. Santa Barbara, California; National Center for Geographic Information and Analysis, Report 89–10

McGranaghan M (1987) Human interface requirements for vehicle navigation aids. *Proceedings*, *Eighth International Symposium on Computer-Assisted Cartography*, Baltimore, MD.: 396–402

Neukirchner E, Zechnall W (1986) EVA-Ein autarkes Ortungs-und Navigationssystem für Landfahrzeuge. Stuttgart, Germany Bosch-Technische Berichte, **8** (1/2): 7–14

Pau L F (ed.) (1990) Mapping and Spatial Modelling for Navigation. Berlin; Springer-Verlag, 357 pp.

Siebold J (1989) Nutzung digitaler Straßendaten für Kfz- Navigationssysteme. Nachrichten aus dem Karten und Vermessungswesen. Frankfurt am Main, Germany, Reihe 1, Heft 103: 113–20

Streeter L A, Vitello D, Wonsiewicz S A (1985) How to tell people where to go: Comparing navigational aids. *International Journal of Man / Machine Studies* **22**: 549–62

Thoone M L G (1987) CARIN, a car information and navigation system. *Philips Technical Review* **43** (11/12)

White M (1985) Building a digital map of the nation for automated vehicle navigation. *Proceedings, Seventh International Symposium on Computer-Assisted Cartography*: 570 (abstract only)

White M (1987) Digital map requirements of vehicle navigation systems. *Proceedings, 8th International Symposium on Computer-Assisted Cartography*: 552–61

Zavoli W B, Honey S K, White M S (1985) A land vehicle navigation system supported by digital map data base. *ADPA Artificial Intelligence and Robotics Symposium*, Nov. 6–7, 1985

Chapter 9

COMPUTER-ASSISTED MAP DESIGN

Wolf D. Rase
Federal Research Institute for Regional Planning, (BFLR), Postfach 200130, D-5300 Bonn 2, Germany

SCOPE

This volume is a mirror of the developments in cartography during the last two decades. Compiled under the auspices of a *cartographic* association, most of the contributions could be published in a volume on geographical information systems (GIS) as well. Some attempts have been made to reflect the changes by providing new definitions of maps and cartography, remote sensing and geographical information systems (among others Hake 1988, Fisher and Lindenberg 1989). The definitions provided there go beyond the scope of the following text. This contribution restricts itself to the *visual display of objects stored in a GIS*. Topics arising from the acquisition, maintenance, administration and transformation of the database, analysis, simulation and other functions of a GIS are not covered, unless they have a direct impact on the display function.

MAPS AND GIS

Until about twenty years ago all maps served two main functions:

—The map as a database: real or imaginary features on, above and below the surface on the earth are stored on an analogue medium (sheet of paper or other material).
—The map as a medium for visual display: spatial objects and associated attributes are transcribed into pictures using graphical or visual variables.

Although strongly intertwined, the two functions are used in maps to different extents. In topographic maps, for example, the storage function has dominated the display function. Thematic maps, in contrast, are designed with an emphasis on the appropriate visual transcription of the objects and their attributes. In the long history of topographical maps the designers were in general successful in finding a sound balance between the two functions. They restricted the number of object types to the ones most important for the main user groups, such as surveying, civil engineering and military. Thus the visual transcription, the map image, remained rather legible even under bad conditions.

In maps for planning purposes, for example, the topographic map is the spatial reference for thematic information. The topographic objects are overlaid by features needed for the planning process. The visual representation of the additional objects in one picture yields a virtually unreadable map (Figure 1). In this example up to nine objects are stacked on top of each other at certain spots. With a more sophisticated use of the graphic variables than in the original map the result can be improved, but is still not satisfying (Lindert 1991).

The introduction of spatial (or geographical) information systems accelerated the development of tools for computer-assisted map design and production. The improvement in the economics of mapping was accompanied by a shift in functionality. The information system relieved the map from the storage function. The entities and their attributes — geometric, topological, temporal, thematic — are encoded in computer-readable form, and stored in the database of the information system (Ganser 1974). It is not necessary anymore to restrict the number of stored object types to ensure a legible map image. In fact the number of object types in planned and working information systems for topographic applications is much higher than in their purely analogue predecessors. The number of attributes and attribute classes for the objects can be increased as well, because the restrictions imposed by the graphic representation are no longer valid. The allegation can be confirmed, for example, by a comparison of the contents of the existing topographic maps in Germany, and the topographic and cartographic information system ATKIS under development (Barwinksi and Brüggemann 1986, AdV 1988).

Until now only thematic maps were subject to cartographic design with respect to the individual map sheet and the purpose of the map. The strict regulations imposed on the visual appearance of topographic maps ruled out any freedom for the individual designer, for example in choosing graphic variables. The regulations have to be maintained for the 'normal' printed edition to guarantee a standard product many users are used to, even if the storage function is transferred to a GIS. When the database in such systems becomes generally available the mapmaker is free to select

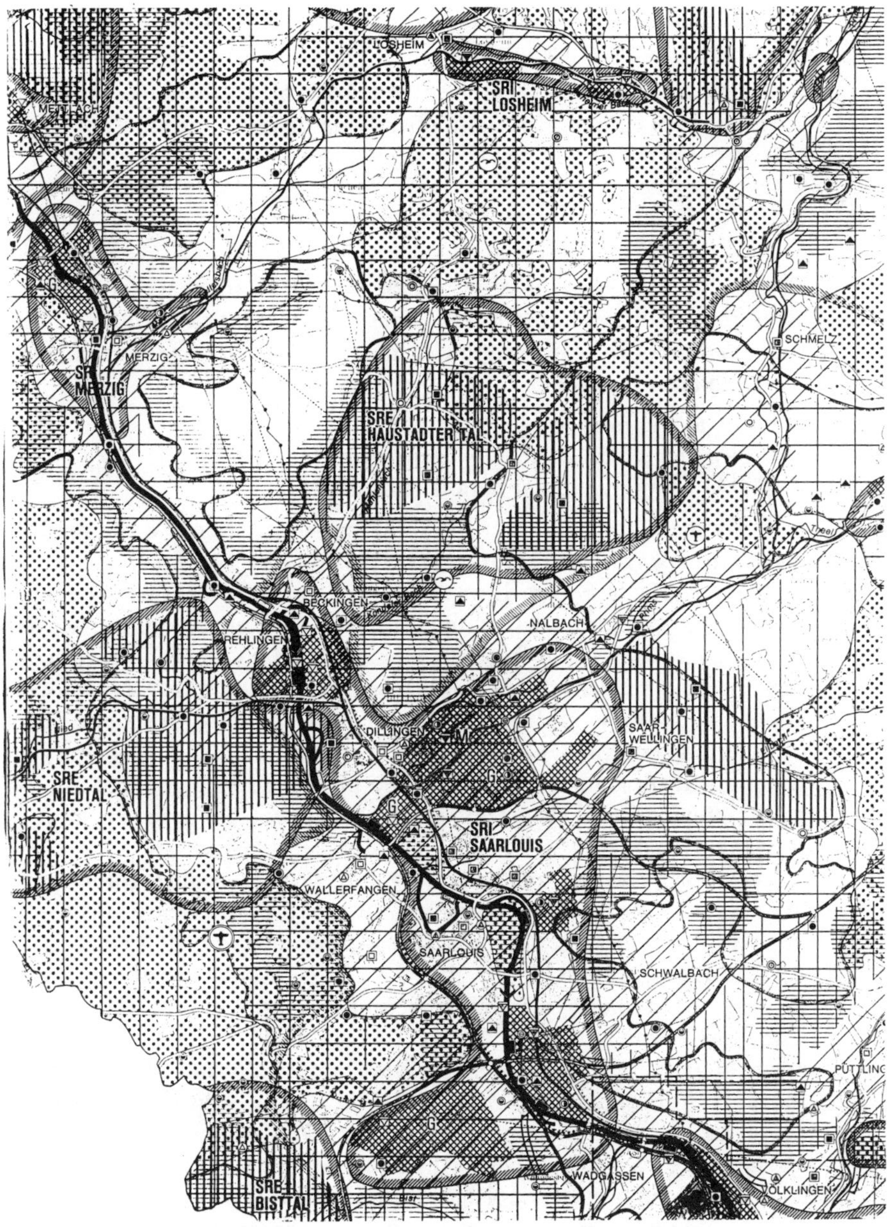

FIGURE 1 Typical map used for spatial planning purposes. (Source: 'Landesentwicklungsplan Umwelt des Saarlandes' 1979) original in colour; the line quality has degraded due to the copy process.

topographic features and to choose how to depict them. The main objectives for the mapmaker are:

—to convey a message as efficiently as possible to the recipient (design);
—to produce the map as economically as possible (production).

MAP DESIGN

Assume for a moment that the use of computers for map production has economical advantages (we will come back to that point later). Efficient transmission of the cartographic message means, among other things, that the recipient (or reader)

—looks at the map at all;
—comprehends the content in general as fast as possible;
—is able to analyse local details if necessary.

Many maps fail to attract the spontaneous attention of the reader, for different reasons. For example, the map contains too much information, such as the one in Figure 1, or the designer has chosen a wrong map type for the objects and the associated attributes. The erroneous application of the visual variables is a frequent mistake. It is a common misconception that the map reader studies the legend first and then uses his knowledge about the symbols and variables applied to analyse the map. The first stage in map comprehension is more or less intuitively to grasp the general situation. If the map is badly designed most readers give up at this point, before they take a closer look. The message the designer intends to transmit is lost.

The appropriate use of the graphical or visual variables (dimensions of the map, size, brightness, colour, form, texture, orientation) is the most important key to successful map design. Bertin (1967, 1974) has provided a systematic framework for the use of the visual variables. But even experienced map designers fail sometimes to apply the variables correctly. The most frequent flaw is the use of colour. Many mapmakers are not aware that the visual variable 'colour' is basically not able to express a sequence, for example to represent data on an ordinal or metric scale. Sequence can only be transcribed by the variables 'size' or 'brightness'.

There are cases, however, where common associations with specific colours can be used to depict ordered data. From blue and red taps we expect cold and warm water, respectively. Consequently shades of blue are used to express data values below (cold), and shades of red (hot) above a certain threshold, usually zero. People with a background in physics tend to associate the sequence of colours along the visible spectrum (the colours of the rainbow) with an ordered sequence. The colours suggested by Peucker (1898) to depict height levels in small scale maps use the associations with the colours of water (shades of blue), arable land and forests (green), and barren land (brown). But these associations are strongly correlated with the educational and cultural background, and are far from common to all map readers.

A simple experiment tests the appropriate application of the visual variables: cover the legend and let someone unfamiliar with the subject guess where the lowest and highest values are located on the map. If the viewer is not able to indicate the extreme values, there is something wrong with the use of the graphic variables, and the map should be redesigned. A lot of maps, even some published in geographic and cartographic journals, will flunk this test.

Concerning the recipient, the map designer is aiming at a moving target. Map readers are usually not one homogeneous group but several groups with different cultural and educational backgrounds, experiences and expectations. The economic advantages of computer-assisted map production allow the designer to optimize the visual image for a specific group, or produce different maps for different groups at marginal additional expense. Variations can be achieved:

—Very simple: varying the class intervals for choropleth maps (Coulson 1987), or the scale factor for proportional symbols.
—More challenging: selecting alternative variables or indices, or choosing another cartographic representation (map type), for example a dot map instead of a choropleth map with area shading or colours.
—More interesting: 'difficult' map types.

The latter possibility demonstrates quite well the new perspectives opened by computer assistance. Many map types described in cartographic handbooks and journals are used very rarely with practical applications, although they have an outstanding capability to transmit the message. They are simply too expensive to draw manually. That is the meaning of 'difficult'. Examples for these maps are, among others;

—Multivariate maps: choropleth maps with two or three variables represented each by a colour or shading; proportional symbol maps where the variables are represented by different visual variables (size, colour, shading, form, orientation).
—Proportional symbols whose manual draughting is time-consuming, such as ellipses, or human-shaped polygons to depict population figures (Figure 2).
—Cartograms: the two dimensions of a map are used as graphic variables; the deviation from Euclidean space has a surprise effect which can be used as an eye-catcher (Figure 3).
—Perspective drawings: bird's-eye views of the topography or buildings; 3D representations are especially useful for an audience not used to reading 2D maps.

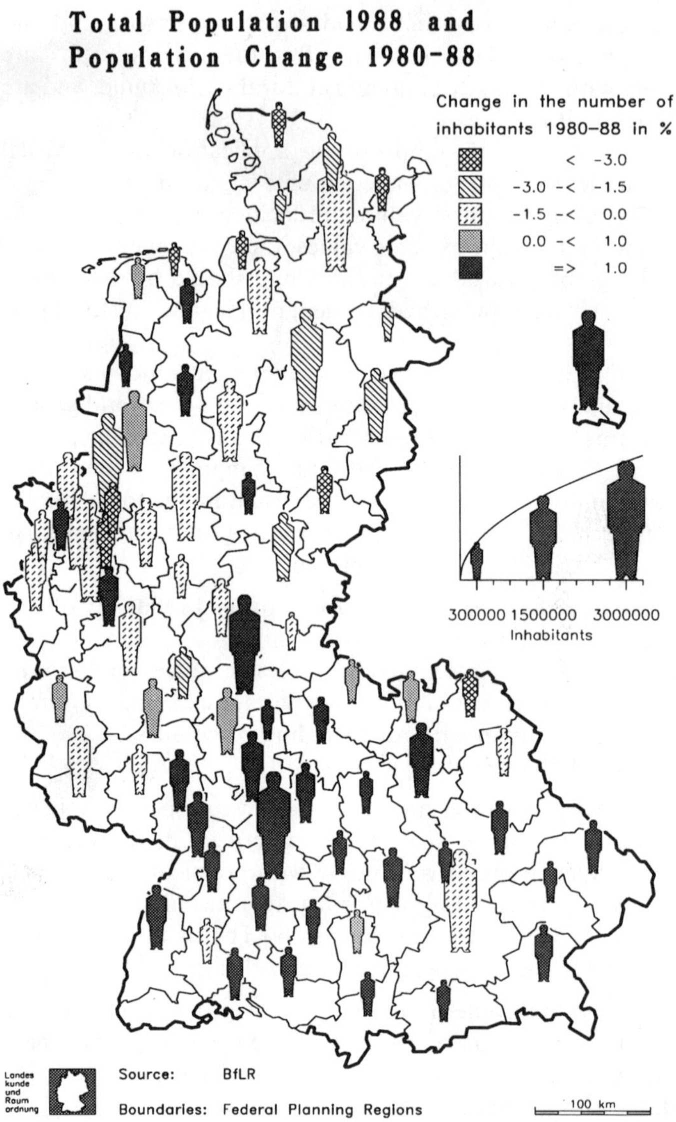

FIGURE 2 'Speaking' proportional symbols. The area of man-like shapes is proportional to the number of inhabitants.

—Computer-generated animation to depict spatial processes over time, to simulate a walk through a planned urban quarter, or a flight over an imaginary landscape, in connection with new display techniques made possible by video and laser technology (Dutton 1979).

Without computer assistance these map types could not be applied on a large scale, because their realization required either a considerable

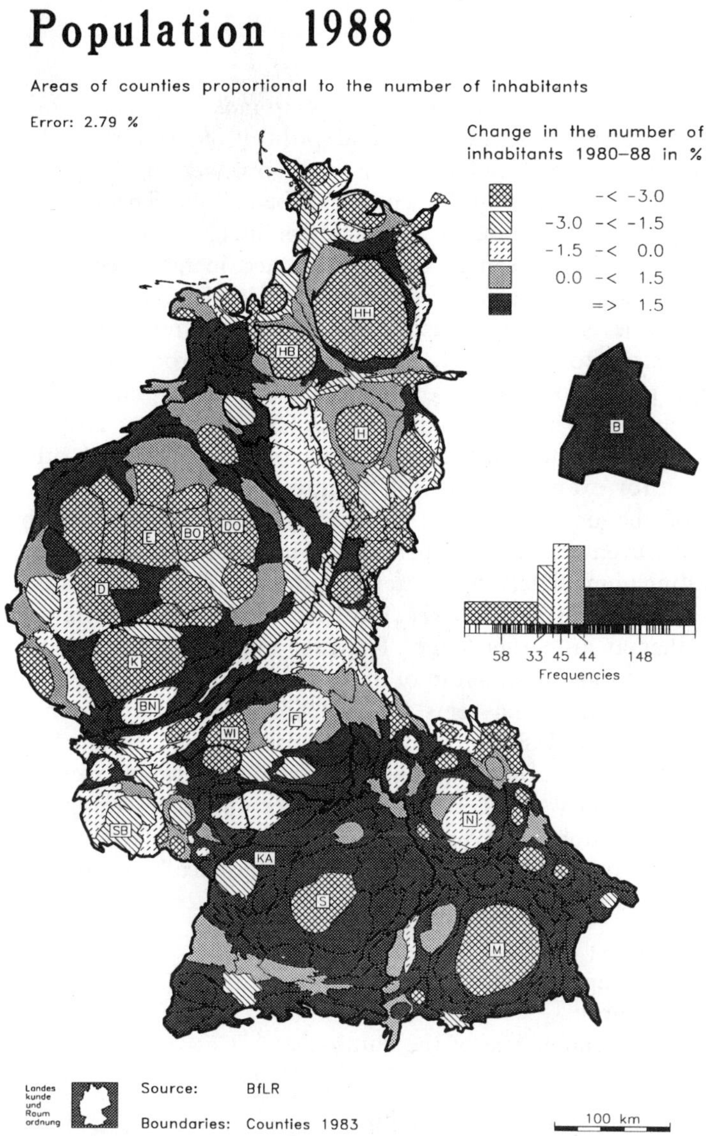

FIGURE 3 Cartogram. The size (area) of the spatial units is made nearly proportional to the number of inhabitants, under preservation of the topological relationships (Source: algorithm of Dougenik *et al.* 1985, slightly modified).

amount of calculations, or the manual construction of the geometric figures was consuming too much time. The computer provides the necessary improvement in economics to make them feasible for routine use.

MAP PRODUCTION

The next step is the implementation of the map-making process for a specific data processing, reproduction and printing environment. For reasons of simplicity I will call this step 'map production'. The term includes the design of the map in the sense described above. The next step after the design is the creation of a permanent image, from a simple hardcopy of the display screen to the map printed in many copies. A consistent production line must be established, which is able to provide different formats, levels of reproduction quality, and number of copies without the necessity for the user to switch to another production line.

A hot topic is the integration of text, tables and graphics in one compound document. Thematic maps are not solitary events. They are usually associated with text which may also contain tables and other graphic images, for example statistical diagrams, or photographs. The components of the document may be created at different times, by different people, even at different locations. The traditional solution is manual mounting immediately before the reproduction and printing steps. In a computer-assisted production line, however, the components have to be fused together at an earlier stage. The user must be able to check, correct and store the complete layout of the document before it goes to the reproduction lab. Software systems for text processing and document preparation based on desktop computers, also known as Desktop Publishing (DTP) systems, offer capabilities of including graphic pictures. Some accept only binary raster images, the better ones also continuous tone images (photographs) and vector graphics, including maps.

The main objective is to produce the map as economically as possible. Several cost factors influence the economics of computer-assisted map-making:

—computers and graphic devices (hardware)
—programs (software)
—acquisition and maintenance of the database
—personnel

Let's have a closer look at these factors.

Hardware

To recover the investment in hardware the number of maps and other graphic outputs should be maximized. To improve the economics the computer system is used for other tasks as well to fill the unavoidable holes. Map-making is seldom a continuous production because the flow of requirements cannot be smoothed out all the time. It is the rule, not the exception, that the computer system serves many other functions of the GIS, for example database management, statistical analysis, documentation tasks in numerous forms, simulation and forecasting, administrative work, or text processing.

The trend in miniaturization of electronic circuits will last for the next few years until the physical limits are reached. Processors become smaller and faster, memory chips double their capacity every two years, and the cost per bit on external storage devices drops continuously. This has some impact on computer graphics:

—Smaller and less expensive computers provide more computing power and storage capacity at lower costs.
—Local and wide area networks allow access to computers and devices at and from distant locations, and the sharing of DP resources, including graphic devices, GIS packages, and cartographic software.
—Mechanical plotters are replaced progressively by raster output devices on several levels of resolution, number of colours, speed, format and cost.
—Faster and more intelligent devices and workstations relieve the main computer of basic graphic functions, such as area fill, segmentation, zoom and pan.

Less expensive hardware will enable individuals and institutions to use data processing systems and apply GIS and computer-assisted mapping where it has not been feasible until now for economic reasons. A few words of caution should be added. The user or buyer must be certain that his computer system is powerful enough both to return a single job and the workload of all users in a reasonable time at minimal cost. The traditional classification into mainframes, minis and micros became rather meaningless for the end user. The familiar buzzwords, such as RISC and CISC, VUPS, MIPS and FLOPS, Whetstones and Dhrystones, and a few others, are more advertisement than a description of the performance (MIPS, for example, has been retranslated into 'Meaningless Information Provided for Sales'). The only possibility of evaluating the speed and throughput of a specific system (hardware and software) is a test with the intended application program.

Software

The investment costs for software are best recovered if as many maps as possible are produced with the same software option or package. The absolute cost of software development will increase because qualified personnel will remain a very scarce, well-paid resource. Besides applying modern software design methods and techniques to facilitate program design, test and maintenance, the software cost per map (relative cost) can be decreased, for example, by

—extending the life cycle of cartographic programs;
—distributing the cost over a larger number of users;
—using existing programs, program modules, and subroutine libraries.

The main prerequisite is the portability of the software. In case of migration to another computer system or the acquisition of a new graphic device the number of necessary modifications to the source code should be minimized, or better avoided at all. The only way to meet this objective is to follow standards adopted by national and international standards organizations. The so-called 'industry standards' not backed by the organizations are only a stopgap and should be avoided.

Programming language

The first essential for portability is the appropriate programming language. The program should compile on a different computer system and/or a different operating system without problems. Of the various programming languages, Fortran still remains the 'lingua franca' of computer graphics, despite its age and obvious deficiencies. One reason is the general availability of Fortran compilers across the whole spectrum of computer architectures and operating systems, from desktops to super-computers. The code generated by modern optimizing compilers is comparable in most cases to the one an experienced assembler programmer would provide. Another reason is the huge investments in Fortran programs and personnel carried out in the past which cannot be abandoned. The present standard Fortran-77 will be superseded gradually by the new standard Fortran-90 which introduces support for data structures and additional flow control statements.

The language C became quite popular among system designers, especially on smaller systems. The UNIX operating system is written in C, thus it is quite natural that C is used to access system functions on UNIX systems. The versions of C and the associated system libraries are not harmonized yet for all computers, but the recent adoption as an ANSI and ISO standard will unify C gradually. Extensions of C, for example C++, support object-oriented programming, which has some potential for GIS and cartography. Pascal is on the decline as a production language. Modula-2 and ADA have gained some popularity, the first in academic environments for its elegance and ease of use, the latter for the economic strength of its originator. Many other languages emerged or resurfaced during the last few years, such as PROLOG or LISP for artificial intelligence applications (Jones and Cook 1989). They are used mostly for prototyping, to evaluate possible solutions, and hardly for production tasks.

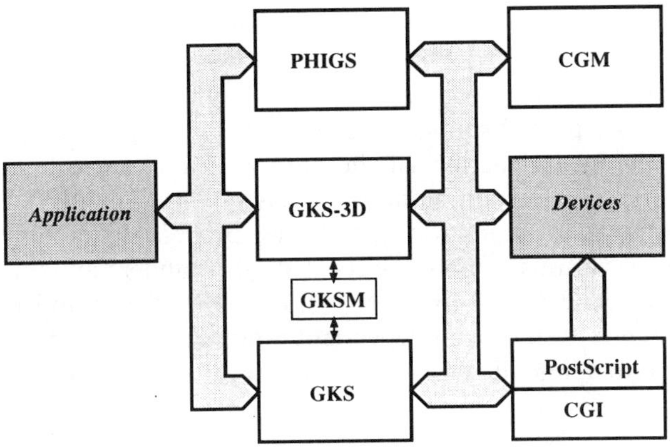

FIGURE 4 The graphic standards and their relations.

Graphic interface

The purpose of the graphic interface is to provide a standardized way to organize output to and input from graphic devices. If a graphic device is exchanged or added, or the program is transported to another system with different graphic devices, no changes to the program should be necessary. The Graphical Kernel System (GKS), (Enderle et al. 1987) is the first graphic standard adopted by the major standards organizations. Although GKS has its shortcomings, it serves cartographic purposes quite well. Other proposals, such as CGI (Computer Graphics Interface) and CGM (Computer Graphics Metafile) are already in the standardization pipeline and will find their applications in mapping (Figure 4). PostScript (Adobe 1985) is rather successful as a control language for hardcopy devices although it is not an adopted standard yet. PHIGS (Programmer's Hierarchical Interactive Graphics System), another standard comparable to GKS, is intended to serve a different constituency, such as highly interactive CAD applications with hierarchical geometric elements. Nevertheless it has been used for cartographic applications (Kraak 1988).

User interface

The keyboard is no longer the only possibility for interaction with application programs. Pop-up and pull-down menus, window systems and pointing devices such as mice and trackballs became familiar tools with desktop computers and workstations. Some standards for user interfaces are emerging; X-Windows and its derivatives and extensions seem to be the most promising ones for general acceptance. The borderline between the user interface and the graphic interface is a little fuzzy. The functions available in GKS, for example, can be used to build user interfaces similar to the ones provided by a window system. Window systems on the other hand are capable of performing graphic input and output. Some implementations of GKS use the window system in the same way as other graphic devices.

In most systems the use of the acoustic channel is limited to the bell of the terminal. More sophisticated sound systems are beginning to emerge which could relieve the visual channel of the high workload typical for cartographic applications in certain situations (Rase 1984). The NEXT workstation, for example, contains a digital signal processor not only capable of producing various sound effects but also understandable human speech. At the input side the picture is less bright. The technical and economical problems of speaker-independent input and processing of spoken words, even with a limited vocabulary, will prevent routine use for the near future.

Operating system functions

Application programs must communicate with the operating system, for example for file access, getting date and time, enquiries about the system environment, and other functions. The programming languages provide a limited assortment of system functions, in most cases only for opening and closing files, or for enquiring file parameters. For sophisticated graphic systems more functions than are available on the language level must be provided, such as memory allocation, timing, support for asynchronous events and parallel processing, and so on. The operating systems provide the functions by subroutine or function calls, but they are system-specific. They have to be modified or eliminated in case of migration.

Portable operating systems such as UNIX seem to be the solution. But contrary to folk wisdom UNIX is not a standard. Neither the command shells nor the system calls nor the utilities including compilers are unified throughout the numerous UNIX versions and their derivatives and clones. A standard (or better a set of standards) known under the name POSIX is supposed to be a cure for the UNIX mess. POSIX defines function calls for interaction with the operating system, for communications, user surfaces, and other areas (Toth 1988). Some major computer manufacturers have committed themselves to implement the POSIX functions in their versions of UNIX and/or their proprietary operating systems. The latter relieves the programmers from the obligation to use UNIX if they want to create a portable program. After all, users need portable applications, not portable operation systems.

Database access

Thematic and geometric data are integrated on thematic maps. The access to the thematic and geometric databases should be standardized to avoid program modifications. Unlike the graphic interface there is no standard for database access. The query language SQL is supported by most database management systems, but SQL is not a standard (Robinson,

Chapter 5). It is at least questionable if the relational model is adequate for geometric data (Frank, Chapter 2). Applications of relational database systems for storage and retrieval of geometric data suffer from severe performance problems (Schek 1986). To be on the safe side an application-specific implementation of the database access is advisable at the moment.

In contrast to the obvious benefits nobody seems to like standards. There are a few good reasons:

—Because the standardization process takes so long, standards are always out of date in relation to technical progress.
—Standards require more resources than a custom-made solution. Because they are a general approach, by definition they impose more overheads on the individual programmer, the computer system and the organization than a solution specially fitted to the application and the environment.

The economical advantages possible by the increased portability and maintainability, together with the faster and less expensive hardware, however, should compensate quite well for the higher overheads in the long run.

Acquisition and Maintenance of the Database

Creating and maintaining the database, both for the geometric and the thematic attributes, is the prerequisite for the visual display of maps. Because it is more related to the GIS than to map making it should be just mentioned here for completeness. Nevertheless it is a crucial issue, especially for the very large databases in national mapping agencies (Haywood, chapter 4).

Personnel

The lack of qualified personnel on all levels, from system designer to operator, from decision maker to user, is the main reason for failures and inefficient operations. The qualification of the personnel has to be brought to the same level as the capabilities of hardware and software. Educational programmes must be established to improve the qualification of the available manpower. Hiring additional personnel with special knowledge might be necessary.

Personnel is the factor where the analogy with classic rationalization — replacement of labour force by machines — is no longer valid. I am not aware of any case where the overall number of employees has been reduced due to the introduction of computer assistance for mapmaking. The impact of data processing on the structure of the cartographic department as well as on the qualification of the individual, however, cannot be denied. The number of employees doing clerical work, such as draughting or mounting, decreases in favour of map designers and DP specialists. It is mandatory nowadays for a professional cartographer to obtain a certain amount of training and experience in data processing. The computer is a tool exactly in the same sense as a ruler, a compass, or an ink pen. The ability to use the cartographic software available at the employee's institution is the minimum requirement. The fluent command of a programming language may not be necessary, but it certainly helps. In any case the cartographer should be able to assess the economic advantages and disadvantages of data processing in a specific situation to select the appropriate solution.

Related to personnel is the problem of uncritical and erroneous application of mapping programs. Cartographic software is easily available: each statistical package has its mapping options, software houses and suppliers offering mapping programs for desktop computers, and a GIS package without mapping facilities would not sell at all. Any person with or without training and knowledge of cartographic principles and rules is able to produce thematic maps. Thus the environment is polluted with meaningless, misleading, unusable and embarrassing graphic output, which hardly deserves the name 'map'. The users cannot be held responsible in all cases. The program designers are to blame, for example, if they provide options contradictory to cartographic rules and experience. Simple checks of the input data and the control commands are often sufficient to avoid the worst mistakes.

The incorporation of cartographic knowledge into programs and systems for mapmaking is a more general approach to cure the disease. Indeed rule-based, knowledge-based or expert systems have been developed and applied for cartographic problems on an experimental basis with promising results (Nickerson 1988, Cromp 1990, de Jong and van der Wel 1990, Muller 1990). Whether artificial intelligence is able to outbalance natural stupidity in general is an open question at this time.

Interdependence of Hardware, Software and 'Brainware'

The 'divide and conquer' approach is considered to be good practice in computer science, as shown in the preceding chapters. Nevertheless hardware, software and 'brainware' have subtle interdependences which should not be neglected. To buy an inexpensive computer is certainly not the most economical solution if the associated software — operating system, compilers, utilities, libraries and application programs — is not adequate for the intended use. A computer with a higher price tag is probably less expensive, because, for example, the application software is readily available. Window systems with menus, usually highly praised for their user-friendliness, are another example of the interdependence. Window systems require additional computer resources, such as CPU power and memory, and are not easy to program. They raise the investment costs for hardware and software, and make the final product more expensive. In most cases the users are willing to pay for the additional comfort. But GIS and cartography are a small market, compared with, let's say, text processing or desktop publishing. For the latter applications the number of program copies sold runs into millions, and the additional expenses for user comfort are marginal.

The users are not a homogeneous group, either. Program designers have to deal with

—Novice users: they need strong user guidance, provided by a menu system, as well as some kind of introduction or tutorial. Their requirements with respect to cartographic options and quality of output are low to medium.

—Casual users: they are familiar with the concepts of the program or system, but do not use it frequently enough to remember all options and parameters. Menus and on-line help are useful functions for avoiding the consultation of printed manuals.

—Power users: they use the programs on a regular basis to produce not just one, but series of maps; due to frequent use they know all options. The cartographic requirements are usually very demanding, including the necessity for a consistent production line from map design to printing.

Most cartographic packages available commercially have a strong front-end with elaborate user interfaces (menus, windows, on-line help). The back-end functions — selection of map types and visual variables, graphic output, support for reproduction and printing — are usually much weaker. Users with languages different from English often search in vain for the special characters in their national alphabets, although the 8-bit ASCII standard supports all Western languages. Unfortunately, most decision-makers endorsing the acquisition of software packages do not belong to the third category.

Closely related to the user categories is the institutional environment. In an educational institution, such as a university, most students are novice users. Desktop computers are able to provide the usually low requirements for data storage and throughput in teaching the principles of GIS and cartography. Research projects need more computer resources and better graphic output devices. The permanent staff doing research belongs to the second or third category. Institutions devoted specifically to spatial research or map production have demands in terms of storage capacity, throughput, output quality, integration of production, etc, which cannot be satisfied economically by desktop computers of the IBM PC/AT variety with the MS-DOS operating system.

It should also be kept in mind that the innovation cycles of hardware, software and personnel differ considerably in time. Processor speeds and memory capacities, constant cost assumed, double every two years. Approximately every five years a novel processor concept emerges. It takes five to ten years before the hardware improvements are reflected by the improvements in software. But ten to twenty years are necessary before our educational system is able to provide the personnel with the adequate qualifications.

PRACTITIONER'S SUMMARY

Our knowledge of map acceptance and map perception is mostly based on intuition, assumptions and personal experience, and to a lesser extent on scientific evidence. The effects of visual variables and cartographic symbols have been investigated systematically (Gill 1988, Wenner 1989, Tainz 1991). Nevertheless we are not sure what really attracts the attention of the reader, how the essence of the map is extracted, how the content is stored in memory, or what makes a specific map type superior to another one under certain conditions. Computer assistance can be a help for the preparation and analysis of experiments to answer such questions.

Computer assistance improves the economics of the design and production of maps, provided the 'critical masses' for economical utilization of hardware, software and the geometric database are present. The tools — computers, graphic devices, programs — are available, but they need better integration and improvement with respect to portability, standardization and user interfaces. The interaction with the user, through either a menu system or command files, is too much inclined towards data processing techniques and procedures, and too little to how cartographers think and work. The situation becomes even worse if, for example, the mapping program is based on a CAD system. The user has to struggle with the cartographic layer at the top, the CAD layer in the middle, and the DP layer at the bottom. My appeal to all designers of user interfaces for GIS and cartographic applications is less Computerese, more Cartospeak!

One of the main bottlenecks for computer-assisted mapping seems to be the qualification of personnel, not only in the DP centre and in software development, but also in the cartography lab and the research department. The educational system begins to provide the necessary qualifications, but educational programmes and on-the-job training have to be established for staff members who have already gone through the educational system. On the other hand, software developers should improve the user interfaces and the manuals in such a manner that the familiarization time with a system can be kept as short as possible.

PROJECTIVE SUMMARY

The state-of-the-art in computer-assisted map production is at the same level as an accurate but rather stupid *cartographic draughtsman*: able to provide excellent artwork from elaborate instructions, but not bright enough to work on its own, or to recognize and correct erroneous commands and logical flaws within the instructions. The mapmaking software has to be promoted to *cartographic assistant*, both to facilitate the design and production process, and to improve the maps and the transmission of their message. Only a few instructions defining the data, the target group and the intended message should be sufficient to create a reasonable map automatically. Of course the proposal can be modified by the human designer if the judgement of the program seems to need correction. To approach that goal we need to know more about

—How maps are perceived (not only at the symbol level), how their content is analysed and stored in the human brain, how they influence the decision process.
—How cartographic knowledge and experience can be collected, encoded and stored, either to be incorporated in algorithms of a cartographic program, or to be used as a database for expert systems.

The development of cartographic software, its integration and standardization, has equal importance, but is more inclined to development and engineering than to research.

REFERENCES AND BIBLIOGRAPHY

Adobe Systems Inc (1985) *PostScript Language Reference Manual.* Reading, Addison-Wesley

AdV (1988) Amtliches Topographisch-Kartographisches Informationssystem (ATKIS). Das Vorhaben der Landesvermessungsverwaltungen zum Aufbau Digitaler Landschaftsmodelle und Digitaler Kartographischer Modelle. Arbeitsgemeinschaft der Vermessungsverwaltungen der Länder der Bundesrepublik Deutschland

Appelt G (1987) Das AdV-Vorhaben 'Amtliches Topographisch-Kartographisches Informationssystem (ATKIS)' — Technische Konzeption. *Nachrichten aus dem Karten- und Vermessungswesen* **I** (99): 15–22

Barwinksi K, Brüggemann H (1986) Development of digital cadastral and topographic maps — requirements, goals and basic concept. *Proceedings of AutoCarto London* 1986 **2** 76–85

Bertin J (1967) *Semiologie graphique*. Paris, Gauthier-Villars

Bertin J (1974) *Graphische Semiologie*. Berlin, de Gruyter

Bjørke J T, Aasgaard R (1990) Cartographic zoom. *Proceedings of the 4th International Symposium on Spatial Data Handling*, Zurich 1990 345–53

Blades M, Spencer C (1986) The implications of psychological theory and methodology for cognitive cartography. *Cartographica* **23** (4): 1–13

Blades M, Spencer C (1987) How do people use maps to navigate through the world? *Cartographica* **24** (3): 64–75

Bollmann J (1985) Theoretische Grundlagen zur Modellierung thematischer Karten. Habilitationsschrift, Fachbereich Geowissenschaften, FU Berlin

Burrough A (1986) *Principles of geographical information systems for land resources assessment*. Oxford

Buttenfield B P, Ganter J H (1990) Visualization and GIS: What should we see? What might we miss? *Proceedings of the 4th Interantional Symposium on Spatial Data Handling*, Zurich 1990: 307–16

Carstensen L W Jr (1986) Hypothesis testing using univariate and bivariate choropleth maps. *The American Cartographer* **13** (3): 231–52

Carstensen L W Jr (1987) A comparison of simple mathematical approaches to the placement of spot symbols. *Cartographica* **24** (3): 46–63

Chrisman N R (1987) Design of geographic information systems based on social and cultural goals. *Photogrammetric Engineering and Remote Sensing* **LIII** (10): 1367–70

Cook A C, Jones C B (1990) A Prolog rule-based system for cartographic name placement. *Computer Graphics Forum* **9** (2): 109–26

Coulson M R C (1987) In the matter of class intervals for choropleth maps: With particular reference to the work of George F. Jenks. *Cartographica* **24** (2) Monograph 37: 16–39

Cromp R F (1990) Knowledge acquisition techniques for spatial reasoning about satellite imagery. *Proceedings of the 4th International Symposium on Spatial Data Handling*, Zurich 1990: 742–51

Dobson M (1985) The future of perceptual cartography. *Cartographica* **22** (2): 27–43

Doerschler J S, Freeman H (1989) An expert system for dense-map name placement. *Proceedings of Autocarto* 9, Baltimore, 1989: 215–24

Dougenik J A, Chrisman N R, Niemeyer D R (1985) An algorithm to construct continuous area cartograms. *Professional Geographer* **37**(11): 75–81

Dutton G (1979) American graph fleeting. A computer-holographic map animation. Harvard USA, Laboratory for Computer Graphics and Spatial Analysis, Harvard University

Ebinger L R, Goulette A M (1989) Automated names placement in a non-interactive environment. *Proceedings of Autocarto* 9, Baltimore, 1989: 205–14

Enderle G, Kansy K, Pfaff G (1987) Computer Graphics Programming. GKS — the graphics standard. Second, revised and enlarged edition. Berlin, Springer-Verlag

Fisher P F, Lindenberg R E (1989) On distinctions among cartography, remote sensing and geographic information systems. *Photogrammetric Engineering and Remote Sensing* **55** (10): 1431–4

Ganser K (1974) Die Aufgabe der Karte im Informationssystem Raumentwicklung. *Kartographische Nachrichten* **24** (5): 169–72

Gill G A (1988) Experiments in the ordered perception of coloured cartographic symbols. *Cartographica* **25** (4): 36–49

Gilmartin P, Shelton E (1989) Choropleth maps on high resolution CRTs/The effects of number of classes and hue on communication. *Cartographica* **26** (2): 40–52

Goodchild M, Gopal S (ed.) (1989) *Accuracy of spatial databases*. London, Taylor & Francis

Guptill S, (ed.) (1990) An enhanced digital line graph design: A feature-based data model for digital spatial data bases that represent geographic phenomenon. US Geological Survey Circular 1048, Reston, Virginia

Hake G (1988) Gedanken zu Form und Inhalt heutiger Karten. *Kartographische Nachrichten* **38** (2): 65–71

Jaakkola T S, Sarjakoski T, Blom T, Laurema M (1990) From satellite data to thematic representation — a knowledge-based system for cartographic visualization. *Proceedings of the 4th International Symposium on Spatial Data Handling*, Zurich 1990: 711–22

Johnson D S, Basoglu U (1989) The use of artificial intelligence in the automated placement of cartographic names. *Proceedings of Autocarto* 9, Baltimore, 1989: 225–30

Jones C B, Cook A C (1989) Rule-based cartographic name placement with Prolog. *Proceedings of Autocarto* 9, Baltimore 1989: 231–40

de Jong W M, van der Wel F J M (1990) Embedded artificial intelligence and spatial data handling — some research and prototyping experiences. *Proceedings of the 4th International Symposium on Spatial Data Handling*, Zurich 1990: 723–31

Kraak M J (1988) *Computer-assisted cartographical three-dimensional techniques*. Utrecht, Delft University Press

Langran G (1989) A review of temporal database research and its use in GIS applications. *International Journal of Geographical Information Systems* **3** (3): 215–32

Lindert B (1991) Einsatz der graphischen Datenverarbeitung bei Planungskarten am Beispiel des Landesentwicklungsplanes Umwelt des Saarlandes (von 1979). Forschungs- und Sitzungsberichte der Akademie für Raumforschung und Landesplanung, No. 185 Hannover 1991

Lloyd R, Searching for map symbols: The cognitive processes. *The American Cartographer* **15** (4): 363–77

McLaren R A (1989) Visualization techniques and applications within GIS. *Proceedings of Autocarto* 9, Baltimore 1989: 5–14

Moellering H (1989) A practical and efficient approach to the stereoscopic display and manipulation of cartographic objects. *Proceedings of Autocarto* 9, Baltimore 1989: 1–4

Muller J-C (1989) Challenges ahead for the mapping profession. *Proceedings of Autocarto* 9, Baltimore, 1989: 675–83

Muller J-C (1990) Rule based generalization: Potentials and impediments. *Proceedings of the 4th International Symposium on Spatial Data Handling*, Zurich 1990: 317–34

Munro P, Hirtle S C (1990) The representation of structure in cognitive maps: An interactive activation model. *Proceedings of the 4th International Symposium on Spatial Data Handling*, Zurich 1990: 732–41

Nickerson B G (1988) Automated cartographic generalization for linear features. *Cartographica* **25** (3): 15–66

Ommer R E, Wood C H (1985) Data, concept and the translation to graphics. *Cartographica* **22** (2): 44–62

Ottosson T (1988) What does it take to read a map? *Cartographica* **25** (4): 28–35

Peterson M P (1987) The mental image in cartographic communication. *The Cartographic Journal* **24** (1): 35–41

Peucker K (1898) Schattenplastik und Farbenplastik. Beiträge zur Geschichte und Theorie der Geländedarstellung. Wien, Artaria-Verlag

Rase W-D (1974) Kartographische Darstellung dynamischer Vorgänge in computergenerierten Filmen. *Kartographische Nachrichten* **24** (6): 210–14

Rase W-D (1984) Computer-controlled input and output of speech: Utilizing the acoustic channel for geometric data acquisition. *Proceedings of the International Symposium on Spatial Data Handling*, Zürich 1984

Rase W-D (1987) The evolution of a graduated symbol software package in a changing graphics environment. *International Journal of Geographical Information Systems* **1** (1): 51–66

Robertson P K (1988) Choosing data representation for the effective visualization of spatial data. *Proceedings of the 3rd International Symposium on Spatial Data Handling*, Sydney 1988: 243–52

Schek H-J. (1986) Datenbanksysteme für die Verwaltung geometrischer Objekte. Informatik-Fachberichte 126, Springer-Verlag: 483–97

Schenk T (1988) Auf dem Weg zu Expertensystemen für die digitale Kartierung. *Bildmessung und Luftbildwesen* **5** (2): 53–64

Sorrel P E (1988) Optimal mapping and the determination of cartographic design principles. *The Cartographic Journal* **25** (2): 128–38

Tainz P (1991) Experimentelle Variation von flächenförmigen Signaturen in Planungskarten. Forschungs- und Sitzungsberichte der Akademie für Raumforschung und Landesplanung, No. 185 Hannover 1991

Toth B (1988) POSIX compliance. *Computer Technology Review* **VII** (4): 1

Uhte A-D (1987) Konzeption einer graphischen Kommunikationsschnittstelle zur Organisation und Weiterverarbeitung geowissenschaftlicher Daten. Dissertation, Fachbereich Geowissenschaften, FU Berlin

Visvalingam M (1989) Cartography, GIS and mapping in perspective. *The Cartographic Journal* **26** (1): 26–32

Wenner A (1989) Welche Aussagen läßt die Blickregistrierung über die Gestaltung aktueller Planungskarten zu? *Wiener Schriften zur Geographie und Kartographie* **3**: 221–32

Wood D, Fels J (1986) Designs on signs: Myth and meaning in maps. *Cartographica* **23** (3): 54–103

Worth C (1989) Problems with experimental research in map design — a case study. *The Cartographic Journal* **26** (2): 148–53

Chapter 10

COGNITIVE CARTOGRAPHY: A NEW HEART FOR A LOST SOUL

David Medyckyj-Scott
Midlands Regional Research Laboratory and Department of Computer Studies, Loughborough University of Technology, Loughborough, UK
and
Christopher Board
Department of Geography, London School of Economics and Political Science, London UK

INTRODUCTION

In the last decade and a half cartographers have increasingly recognized the importance of cognitive processes in the study of cartographic communication and map use. In this chapter we describe the growth of cognitive cartography and what it has to offer in furthering our understanding of cartography. Although the main focus is on map use rather than map design, we shall also indicate why and where we think cognitive studies will have an increasingly important role to play in cartography, even in map design.

From the early 1970s, geographers and psychologists began to pull together work on cognitive (or mental) maps and cognitive mapping. Cartographers, moving towards professional independence from geography as a discipline, recognized the importance of this emerging field rather later. More recently still a new field called cognitive mapping has arisen in operations research where the cognitive maps are essentially diagrams linking elements, eg in problem solving, in a structured fashion (Eden 1988). It has little in common with the subject of this paper and will be excluded from it.

CARTOGRAPHY AS A CRAFT

According to Long and Dowell (1989) a discipline whose knowledge is based on the practice of its craft often fails to be fully effective. As a craft, cartography solves the problems of map design by carrying out design and evaluation, the latter generally informally. These practices are supported by knowledge typically in the form of rules or heuristics, the latter implicit and informal. This knowledge is acquired by practice and example, in other words, by experience. Cartography as a craft is therefore maintained by the practice itself. Heuristics are generated for a purpose, tried out for their adequacy and then retained or discarded. Therefore the costs incurred are few and in this sense the craft of cartography is effective.

Consideration of cartography as a craft based discipline however suggests that it fails to be effective for a number of reasons. Knowledge possessed by cartography as a craft based discipline is not operational. Because the knowledge is either informal or implicit, it cannot be directly applied by those who are not associated with the heuristics or exposed to their use. If the heuristics are informal, they can be applied only with the help of guidance from a successful cartographer (or, in some cases, by additional, un-validated, reasoning by an individual). If the heuristics are implicit in practice, they can be applied only by the means of examples.

If craft knowledge is not operational, it is also unlikely to be testable (Long and Dowell 1989). Cartography as a craft based discipline again fails to be effective because there is no guarantee that the heuristics applied will have the intended consequences since guarantees cannot be provided if testing is not possible. Moreover, if the craft knowledge is not testable, it is unlikely to be generalizable: there is an absence of generality in the knowledge and thus no general theory of map design can be derived. There can be no guarantee that the knowledge will be appropriate and effective in instances other than those in which it was derived or generated. In the absence of a general theory, each map design problem has to be approached anew, possible heuristics applied and tried out.

Thus, although the knowledge when applied by practice is very often effective, the actual discipline of cartography from this viewpoint is ineffective because it generally cannot be used to derive a general theory of map design. It is ineffective because its knowledge is not operational (except in practice itself), nor generalizable, nor guaranteed to achieve its intended effect, except through the continued success of its practice and its continued use by skilled cartographers.

Recognizing these weaknesses, cartographers have looked to a more scientific approach to their discipline. This has been through a search for more general principles initially from a design perspective, but more recently from a psychological viewpoint. (Compare Robinson (1952) and Bertin (1967) with Raisz (1938) and Monkhouse and Wilkinson (1952).)

HISTORICAL ASPECTS

Cartography as a Psychological Science

It is nearly 40 years since Robinson published '*The Look of Maps*'. One of its central themes was that map design was a subject suitable for scientific analysis. Thus, from a cartographic viewpoint, the book was decisive in initiating a more scientific approach to map design. As Petchenik (1983) points out, Robinson set a new goal for cartography. Maps had to be 'correct' graphically as well as portraying information correctly. This demanded a greater interest in the map user, leading to 'the development of design principles based on objective visual tests, experience and logic; the pursuit of research in the physiological and psychological effects of colour, and investigations in perceptibility and readability in typography ...' (Robinson 1952: 13). Thus was established the link between cartography and psychology which persists today.

This relationship arose from cartographers searching for new ideas and approaches in other disciplines. In the 1950s several new ways to approach map design were promising. Some, such as psychology, physiology and optical studies were well established. Others such as Human Factors (Ergonomics) had only recently emerged as a result of knowledge needs during the Second World War. Following the publication of '*The Look of Maps*' cartographers began to conduct experiments in map design using methodologies taken predominantly from psychology.

Cartography, Neobehaviourism and Psychophysical Studies

At this time the dominant paradigm within psychology was neo-behaviourism. For about forty years neobehaviourism (and its parent behaviourism) had enjoyed an unprecedented degree of allegiance among American and to a lesser extent British psychologists. In their purest form behaviourism and neobehaviourism were based on the notion that psychology could be scientific only if its focus was on what was observable — environmental stimuli and behavioural responses. As a consequence, early behaviourists made considerable theoretical use of stimuli–response connections to develop S–R Theory. The weak point of behaviourism was that it effectively ignored the existence of mental life and thought processes.

Neobehavioral psychophysical studies modified the hard line behaviourist position, focusing on experimental research on relationships between physical events in the external world (stimuli) and the corresponding processes of perception, observation and judgement in the brain. They were therefore seen as relevant to the study of map design. In its narrow and traditional sense, psychophysics stemmed from the work of G.T.

Fechner who realized that the difference in intensity of sensation appeared to be equivalent to the logarithm of the change in stimuli strength. In 1957 this was refined by Stevens' so-called 'Power Law', which stated that 'on all continua governed by the power law, a constant percentage change in the stimuli produces a constant percentage change in the sensed effect' (Stevens 1975:16).

Ignoring the mainstream behavioural work, cartographers took rapidly to the ideas of psychophysical studies and with them gained their first taste of experimental studies. Stevens' power law was eagerly adopted and applied in tests of a large number of map symbols. Specifically cartographers attempted to discover experimentally the mathematical relationship between the actual and perceived characteristics (usually size) of map symbols. Initial studies soon showed that subjects underestimated symbol dimensions. The first true cartographic applications of the psychophysical approach were undertaken, independently of each other, by Williams (1956) and Flannery (1956). These were followed by a decade or more of experiments in which quantitative symbols, lettering and grey scales, were subjected to psychophysical testing — Ekman et al. (1961), Robinson (1967), Meihoeffer (1969), McCleary (1970) and Kimerling (1975).

In the late 1960s and earlier 1970s the emphasis on psychophysical studies began to be widely criticized (Petchenik, 1975). Most psychophysical studies made use of highly simplified maps with little base information or merely symbols within a frame. By doing this it was intended that the subjects taking part in the experiments would be presented with as few distracting elements as possible while they would regard the test conditions as resembling a normal map reading situation. This restricted the types of questions that could be investigated and the application of the results to genuine but more complex map reading situations. Further criticisms of this type were made by Cox (1976), Chang (1977, 1980) and Castner (1983). Thus Gilmartin realised that all this research threw little light on how people read maps in the sense of learning, remembering and making use of them (Gilmartin 1981: 11).

By the earlier 1970s, although psychophysical studies continued, many cartographers slowly changed the direction of their research. Although there is no direct evidence for it, it is evident that there was also a loss of direction, due perhaps to the lack of a theoretical framework. Surprisingly, in searching for some kind of map theory, cartographers did not look to psychology but instead focused on information theory and communication theory. The reason can be found in contemporary events in psychology.

The Growth of Cognitive Psychology

In the 1960s psychology was in the process of a massive paradigm revolution. This began when the conventional objectives of the learning theorists were confronted with proposals that the formation of associations between stimuli and responses were not the only kind of learning and that humans often learn in other ways (Lachman et al. 1979).

Consequently psychology came to be predominantly a study of mental processes: perception, memory, attention, pattern recognition, problem solving, language and cognitive development. Psychologists, like cartographers, looked to new areas from which to borrow concepts, theoretical formulations and terms and units of analysis. Those investigated were primarily communication engineering and information theory, linguistics and computer science. For example, efforts to write computer programs to simulate and model cognitive processes alerted psychologists to the multifarious ways in which humans use their knowledge and suggested further hypotheses. The result of this activity was the emergence of a new paradigm which came to be known as 'Cognitive Science'.

Influence on Cartographic Research

When any subject undergoes a scientific revolution, acceptance and adoption of the new paradigm does not occur simultaneously, nor are all convinced of the need for change. Such was the case with psychology in the 1960s. Cartographers, looking to psychology for inspiration, would have been confused and would have found little help there for defining a theoretical framework for cartography. As a result, several cartographers independently began to look to communication theory and information theory to investigate their relevance to the map context. (See the summary by Board (1981).) But despite all the theoretical activity that occurred, few attempts were made to link the theory with practice, although Board (1967), Wood (1968) and Morrison (1974, 1975) examined this relationship.

The earliest models of cartographic communication adopted a relatively global outlook on the processes, but later ones were concerned with local (sub-process) analysis of the communication system. The result of the latter emphasis was that cartographers began to think about the actual processes involved in the model rather than treat each stage as a black (or at best, grey) box. Interest in two of these areas was particularly important. The first was the way in which the different elements in the map were 'put together' to communicate most efficiently the message that the map author wished to send, and secondly, the perceptual and cognitive processes used during the decoding stage of the message by the map user, ie the processes of map reading.

The former had two results. The idea of the map as a cartographic language, into which the assembled content (of the map) had been converted, appealed to many. This is sometimes called encoding, but frequently it is assumed to have some parallel with natural language and its structure (Head 1984). Drawing on linguistics, semiotics, and psychological research on human information processing, Head argued that the idea of a natural language of cartography has sufficient merit to be considered as a basic, general theory of cartography. Others have chosen to interpret the process of cartographic communication primarily in terms of semiology (Bertin (1967) and Freitag (1971)). This concept of a cartographic language has not been universally accepted. Robinson and Petchenik (1976) and Keates (1982) have produced arguments against it. To date there has been little research into this aspect of the cartographic communication model.

An information processing view

Empirical work continued, with the objective of seeing how maps could be designed to be easily and efficiently understood and were aimed at improving the communication efficiency of the map (Dent 1972, Head 1972, Cuff 1973, Jenks 1975). However by increasingly treating the psychology of the map user this type of research has become indistinguishable from the work on map reading processes.

During the 1970s work began on the processes used by the map reader for decoding map information. This and the sterility of psychophysical studies, led cartographers to realize that it was vital to be specific about how maps were used, what map-using tasks were being performed and how mapped information was being extracted (Hopkin and Taylor 1979). The goal was to identify the ways in which the map and its information were used and how its design might influence that use. There were two distinct but related reactions to this from cartographers. Some returned to the psychological literature for a more extensive review of the current research and how it might be related to map perception (Wood 1972, Robinson and Petchenik 1976, Dobson 1979, Hopkin and Taylor 1979). Wood (1972) stressed that it might be the abilities of the reader, and not the efficiency of the map, that limit the chances of successful cartographic communication. Others were attracted to the ways in which maps were being used and the tasks performed upon them (Morrison 1976, Olson 1976, Board 1978). Surprisingly few cartographers have extended the theoretical work on tasks or how the more basic tasks are combined into more complex ones. However the psychological literature provides guidance here, either from the work done specifically on tasks, (Shaw 1963, Fergusson 1966, Hackman 1969, Wood 1986), or from the research conducted on problem solving (Newell and Simon 1972, Anderson 1980, Simon 1985).

One of the early outcomes of the cognitive approach to psychology was the attempt by psychologists to produce flowcharts designed to reveal the internal functioning of human beings. Such early information processing models (Waugh and Norman 1965, Atkinson and Shiffrin 1968), represented mental operations as a linear sequence of stages with only a very limited amount of interaction, mostly unidirectional, between stages. These simple models were superseded by systems with multi-directional connections between their components. Cartographers such as Wood (1972) and Dobson (1979) were attracted to this approach, and to the models, as a basis for explaining map perception. It was natural that cartographers would come into contact with the information processing approach, then the dominant paradigm in psychology. Moreover the whole concept of information processing appealed to cartographers, who saw an analogy between it and the communication of map information. The models were also very attractive, because the early ones were simply understood by non-psychologists and because they provided 'confirmation for principles of visual logic long practised by cartographers in contact with their user population' (Dobson 1979: 19). Such research was readily, perhaps too readily, incorporated into cartographic theory and applied to explanations of the map–map reader interface.

The growth of interest in the map–map reader interface did not take off until the middle 1970s, while experimental work within cartography continued to use psychophysical methods. Those dissatisfied with this approach turned to visual search experiments (Phillips et al. 1975; Wheate 1978); eye-movement recordings (De Lucia 1976, Dobson 1977, 1979, Castner and Lywood 1978, Phillips et al. 1978), and studies of map complexity (Jenks and Caspall 1971, Olson 1975, Eastman 1977), in order, they believed, to gain a better understanding of the nature of maps.

The 1970s saw a resurgence of the influence of psychology on cartographic research as the ideas of information processing became better known. Thus cartographers were introduced to cognitive processes but the cognitive element was restricted by the focus on perception studies. Two encouraging developments came out of such cartographic research. Cartographers attempted to understand and measure the effect of the entire map design on the map user's performance. Greater emphasis was placed on defining the tasks the user carried out and on attempting to incorporate them into the experimental design as a means of evaluating the map. After 1976 the use of tasks in experimental evaluations of maps increased. For almost the first time, tasks provided a means by which the map–map reader interface could be investigated on topographic maps as well as simpler thematic maps. Oddly this conceptual framework has still received little attention.

AREAS OF COGNITIVE PSYCHOLOGY RELEVANT TO CARTOGRAPHY

Cognition: A Definition

Cognition is defined by Neisser (1967: 4) as '...all the processes by which the sensory input is transformed, reduced, elaborated, stored, recovered and used...such terms as sensation, perception, imagery, retention, recall, problem solving and thinking, among others, refer to hypothetical stages or aspects of cognition'. Cognitive research methodology involves laboratory experiments, modelling and simulation techniques.

Visual Cognition

Most work by psychologists has concentrated on particular operations, such as the reaction of retinal or cortical cells to stimulation and the capacity of Short Term Memory (STM). Research on the mechanisms of visual perception has led to greater understanding of how visual processing occurs. Recently there has been renewed effort to understand the perception of real world scenes, in an attempt to achieve more ecological validity. Visual Cognition, as this research has become known, is an expanding area of psychological research and includes within its scope visual recognition and visual imagery (Pinker 1985: 4) and visual routines (Ullman 1985).

Visual memory

Current psychological thinking views memory as a symbolic, semantic network of nodes and links in what can be viewed as a hierarchical structure. The structure is multidimensional with numerous strands joining related concepts in different subject areas. Originally memory was believed to consist only of verbal codes but alternatives have now been proposed (Paivio 1969, Anderson 1978, Kosslyn et al. 1979).

For example Anderson's Tri-code Theory proposes that humans process three kinds of information: visual–spatial, verbal–sequential, and abstract–propositional; and that memory consists of three separate codes with the potential for interrelations among them. Considerable effort has been expended on investigating visual Short Term Memory (STM) (eg Phillips 1983) and visual Long Term Memory (LTM) (eg Kosslyn et al. 1979). While it has been generally agreed that the experiments on visual STM show strong support for the view that visual cognition has components and properties analogous to, but different from, those of STM and LTM in verbal cognition, the properties and form of LTM are still hotly disputed, as is the relationship between image or visual representations and other forms of knowledge representations.

Knowledge Representation

In order for the map user to use his knowledge it must be represented in memory in some form. A number of representational systems have been hypothesized. The most popular today belong to one of three system categories: propositionally based; procedural representational; analogical representational (Rumelhart and Norman 1985).

Some psychologists refer to experts' representation of knowledge being stored in the form of schemata which are 'data structures for representing the generic concepts stored in memory. They exist for generalized concepts underlying objects, situations, events, sequences of events, actions and sequences of actions' (Rumelhart and Ortony 1977: 121). Schemata are, therefore, cognitive structures of organized prior knowledge, abstracted from experience and becoming more integrated and flexible, through accretion, tuning and schema restructuring. They guide the processing of new information and the retrieval of stored information. Schema (plural, schemata) is often used as a generic term to refer to bodies of knowledge and their representation in the mind. However, other labels can be found, sometimes reflecting slightly different interpretations of the concept. Minsky (1975) uses the term 'frame', while Schank and Abelson (1977) employ the terms 'scripts' and 'plans'.

Other Areas

Probably the most important of these for understanding map reading are spatial cognition (Pick and Acredolo 1983) and spatial linguistics (Herskovits 1987); but the work on cognitive development and learning (see Sophian 1984), and problem solving (see Greeno and Simon 1988) also have something to offer.

The research referred to above offers the potential for furthering our knowledge about how the map and the real world are fused together by the user (in the presence of the map or away from it). Furthermore the suitability of the psychological ideas incorporated into cartography over the years needs to be reconsidered. Our models of the perceptual processes need to be re-thought to include, for example, the properties of visual STM and the various representations in LTM.

COGNITIVE CARTOGRAPHIC RESEARCH: A NEW APPROACH

Fifteen years ago Petchenik critically reviewed the previous ten to fifteen years of perceptual research by cartographers in the framework of behavioural psychology. This research did not amount to much for those confronting the problems of map making. 'No whole theory or set of principles, greater than the sum of the component parts, has emerged. Similarly, analytical attempts to deal with the notion of map reading have not led to any theoretical structures from which principles that would assist in the details of map design can be deduced. Clearly map reading is more than just the accumulation of a number of simple perceptual considerations of symbol size or value' (Petchenik 1975: 183).

Robinson and Petchenik developed this criticism further by arguing that not enough attention had been paid to the primary element in the mapping process, 'namely, the cognitive characteristics of the mapper as both a cartographer and a percipient' (Robinson and Petchenik 1976: 108). According to Robinson and Petchenik map research 'must deal with more than just the physical character of the map. It must probe the characteristics of human beings as they see and know, spatially, and how they use maps to understand and communicate this knowledge. The emphasis must shift from the map as a static graphic display to the cognition and perceptual activities of the individuals who interact with maps, namely, those mappers who are map makers or map percipients' (Robinson and Petchenik 1976: 21).

Although writers had mentioned the relevance of cognition to cartographic research (Gamezo and Rubakhin 1964, Aslanikashvili 1968, 1974, Wood 1968, Dent 1972), it was the series of essays by Robinson and Petchenik (1976) that led cartographers to realize the importance of cognition to cartographic issues and so, subsequently, brought them back to the 'state of the art' work in psychology. What then is distinctive about cognitive research?

The distinctiveness of cognitive research lies in the way it views people. Previously, they were viewed as passive receivers and stores of information, but to cognitive psychologists, they are seen as active individuals who can filter information, act selectively, organize experience and create. Many cartographers now take a similar view. Robinson and Petchenik (1976) viewed seeing as an active process — a transaction between the individual and his environment, in which the viewer creates some kind of order from the essentially disordered array that is the visual stimulus.

What is important to cartographic research then is not just the map, its message and the uses to which it is put, but also the knowledge that the user brings to the map-using environment, his personal, meaningful, mental, spatial concepts of the world; his past experiences, decisions made, and the goals set before and during the act of map reading. Perception is not just about processes such as detection, discrimination and identification but also involves interpretation, an active search for meaningful clues by the projection of aspects of our cognitive schemata onto the map data. Perception and knowledge are part of a continuum, rather than separate entities which interact with one another. Because knowledge and past experience is so important to the acquisition of meaning, what may be learned from the map cannot be an exact duplicate, or even strictly a subset, of the original stimuli. For the cartographer who designed the map, and the individuals who read the map, vary in the type and amount of experience, and the size of the knowledge domain they possess.

Map use can, therefore, be viewed as a two-way process between the map and the map reader. This suggests a distinct area of research in cartography although it is a natural outgrowth of work on map tasks, map use and perceptual studies. Cognitive cartographic research is concerned with the active processing of the map, which is influenced not only by the map but also by the cognitive structures and processes such as memory, thinking, imagery, motivation and attention. It looks to the development and testing of theories dealing with the map user's interaction with the map, thus culminating in a different set of problems to be solved rather than a different method of solving old problems (Olson 1979).

Cartographers are presented in the 1990s with two possible directions for research: to investigate only the perceptual problems of map design; or to emulate psychologists and study the higher mental processes of map reading. Keates (1982: 60) is clearly convinced that 'if map use is a cognitive act, designed to acquire and integrate new knowledge on the basis of the map user's needs, interests and existing knowledge, then progress is more likely to be achieved by studying and observing people engaged in real map-using tasks'. If cartographers choose the latter route they cannot expect a quick dividend to advance our knowledge of map design, but there is some prospect that ultimately cartographers will be able to learn more about the map user and hence to provide solutions to the problem of map design.

RESEARCH IN COGNITIVE CARTOGRAPHY

Since the 1970s there has been a growing volume of research which can be called cognitive cartography. Initially it ranged from study of mental constructs emerging from the interaction of cognitive processes and the map (Olson 1979); through the role of 'Visual Cognition' in the comprehension of maps (Robinson and Petchenik 1976); to those cognitive aspects of map reading influenced by differences in people's experience and cultural background (Guelke 1979).

We can recognize two strands of cognitive research: one devoted to particular cognitive phenomena and the other describing map readers' strategies in particular tasks. Because most of the research has been experiment driven, attempts to explain the processes and their interrelationships have been limited. Modelling these processes, important in cognitive research, has rarely been attempted (McNamara (1986) is the exception), but this may reflect a clear difference in focus between the psychologist and the cartographer. Bartz (1970), Eastman (1977) and Peterson (1978) provide examples of early experimentation including a cognitive aspect. More recent studies have concerned themselves with higher order map-reading tasks (Board 1984).

The complexity of the cognitive processes employed during map reading varies with the difficulty of the map reading task and the cognitive task. A cognitive task involves the mental processes in which information is perceived, selected, compared, stored and recalled (Blades and Spencer 1986*b*). The information may be from a source external to the individual (a map), or from an internal source (from memory), or more generally there is interaction of information from both these sources. Some tasks may be simple and of low complexity, others difficult and extremely complex, as in the visualization of terrain. Here, as well as interpreting the map, an image has to be constructed, transformed, possibly scaled, rotated or seen from a different viewpoint. With experience the difficulty of tasks may be reduced, but the complexity of the mental processes is unlikely to change.

Cognitive Phenomena

Individual differences

Once cartographers began to be interested in cognition, they soon became aware of the great variation in individuals' performances in experiments. Such variation occurs in the inherent characteristics of the individual: experience and knowledge gained from previous contacts with maps and with the environment; different strategies adopted to tackle a task; and characteristics acquired from being in their social and cultural milieu.

One topic concentrates on differences occurring in a single individual. Why does someone perform a task better or worse than others? Does an individual possess greater or lesser spatial ability, visualization ability, visual memory, verbal or visually dominated mental processes or even map-reading skills? Can such differences be reduced by training? Another topic examines the differences between individuals and groups, such as the difference between expert and non-expert map users or between males and females (Gilmartin and Patton 1984). Do occupation, intelligence, age and the nature of the individual's home territory create individual differences and why? Relatively few studies of individual differences have been done by cartographers; notable studies include the following: Muehrcke (1973), Cromie (1977), Thorndyke and Stasz (1980), Sholl and Egeth (1982) and Eastman and Castner (1983). These have tried to interpret the individual differences by correlating various variables of the subject used in the experiment against their responses.

Mental representations

In studying mental representations in map reading, Scott (1987) and Eastman (1985*a*) have investigated the role of structural descriptions (Minsky 1975, Palmer 1975). Head (1984) sees mental representations as propositional structures consisting of subject and predicate, sometimes subdivided into relation and object. Head uses this 'deep structure' as the basis for proposing a natural language and linguistic theory of map reading. Steinke and Lloyd (1983), Lloyd and Steinke (1985*a*, *b*), Peterson (1985) and Scott (1987) are among a number of workers to investigate the role of mental imagery in map reading. Scott has attempted to demonstrate the relationship between mental imagery and visualization tasks, pointing out that imagery may be involved in more tasks than cartographers realize: for example in the recognition and orientation of maps (Rock, 1974, Cerney and Wilson 1976, Steinke and Lloyd 1983); comparing graduated map symbols (Kerst and Howard, 1978, Moyer et al. 1978) for cognitive biases and Lloyd (1982) for relationship to map context; and finally route choice (Shepard and Hurwitz 1985). Arguably the most comprehensive, cognitively-based description of this topic is provided by Scott (1987).

The organization of map detail has interested a number of cartographers following the work of Chase and Simon (1973) who developed a model for the memorization of chessboard positions. They proposed that players recall situations through a series of 'chunks' or sections of the board, each chunk containing pieces involved in a common or recognizable structure. Chunks are mentally linked together, one chunk leading to the next during recall, or as one attempts to memorize it, until the whole situation has been recalled. Evidence for chunking during map reading comes from Wood and Kinnear (1984) while Gatrell (1974) demonstrated that map images were structured within 'cognitive spatial chunks' established by such features as coastlines and rivers, and that subjects recalled the map by distinct geographic regions. Eastman (1982, 1985*b*) has carried out experiments on the organization and structuring of map information. From the results, he postulated a 'Hierarchical Model of Map Learning' in which the spatially arrayed material on a map is partitioned into a series of chunks linked into a hierarchical network of spatial relations. Important research into the role of schemata in map reading has been done by Williamson (1987).

Map Reading Strategies

High and low level tasks

Several researchers have investigated more complex higher level tasks (Cromie 1977, Wheate 1978, Eley 1981, 1983*a*, Cross et al. 1982, Griffin

1983, Simitus and Barsam 1983, Wood and Kinnear 1984), but our knowledge of such tasks is still limited, for example with respect to visualization.

Visualization, the process of creation and manipulation of mental images, has been a process rarely explored, but often discussed. Few have defined the term, and even fewer have tried to describe the perceptual and cognitive processes involved. Board (1978: 2) describes visualization as 'a peculiar geographical skill demanding a synoptic view of the distribution(s) portrayed on the map(s). It is based on the map reader's ability to synthesize and relate to one another the form and pattern of individual features'. Visualization is of interest to many because it is so little understood, as Chang et al. (1985) recognised. Visualization intrigues others because it involves an attempt to understand several seemingly complex cognitive processes. However very little is yet known about the cognitive processes and mental structures used during visualization or how individual differences affect performance. One promising direction for research would be to develop much needed cognitive models of map tasks such as visualization but little attempt has been made to do this except by Scott (1987).

Research on lower level tasks from a cognitive perspective is also limited but includes some on the recall and recognition of regional shapes (Shepard and Chipman 1970, Prince 1974, Lloyd and Steinke 1984, Scott 1987), comparisons of point symbols (Lloyd and Steinke 1985*a*, 1985*b*), and discriminating direction from maps (Shepard and Hurwitz 1985).

Remembering and recalling map information

People often memorize part or all of a map in order to perform such tasks as selecting routes or estimating distances between points. Shimron (1978) has reported on how individuals learn about the location of features from textual and map information while Gilhooly et al. (1988) have demonstrated that skilled map readers remember map detail better than novices do when the maps are complex. Thorndyke and Stasz (1980) attempted to study the cognitive process involved and to determine the learning strategies used when studying a map in order to understand why individuals' information acquisition rates varied. They asked their subjects to learn a number of maps until they could recall them with no errors (an unusual task rarely performed by a normal map user). From verbal descriptions, written questionnaires and drawings they were able to categorize several learning procedures invoked during map reading. Good learners used primarily spatial encoding procedures such as imagery, while poor learners relied more heavily on verbal learning procedures.

Children's Use of Maps

Understanding how children use and read maps (Presson 1982, Freundschuh 1987) and how the form and nature of this use changes as they develop (Bluestein and Acredolo 1979) has absorbed considerable research effort. It has focused on the development of understanding that maps are symbolic representations of real space; the development of map concepts in children (Liben and Downs 1989); how young children acquire information from maps and represent that spatial information (Uttal and Wellman 1989) and how they navigate using maps (Blades and Spencer 1987*a*, *b*). From his long term interest in the mapping abilities of quite young children, Blaut has proposed a theory of mapping behaviour (Blaut 1987).

PROBLEMS IN THE COGNITIVE APPROACH

There are three main problems affecting the cognitive approach to the study of map use. While weakness in experimentation has been inherited from the pre-cognitive era, the others are *sui generis*.

Weakness in Experimentation

A cartographic experiment starts with the formulation of the problem in the form of a question, which is often difficult to define, especially in cognitive studies of map reading. This was more true when map making was employed as the paradigm for defining the elements of map reading and hence the hypothesis formulated (Olson 1979). This may have been due to the influence of models of cartographic communication which stressed the relationship between the activities of the map maker and those of the map user. Some have argued that this 'communication paradigm' should be rejected for a 'cognitive paradigm', but we require both the strengths of the cognitive paradigm combined with the direction of the communication paradigm. Each possesses a valuable and distinctive methodology. However, although the adoption of the scientific method and experimentation in cartographic research makes the discipline methodologically rigorous, research is too often flawed. Cartographers rarely have a background in experimental design so that a familiarity with the relevant psychological literature (Federer 1955, Edwards 1970, McGuigan 1978) is essential to improve the quality of experiments.

Unfortunately there has been little replication of experimental work, a fundamental scientific criterion for evaluating the validity of conclusions. This flaw may stem from the small pool of active researchers, a host of unanswered questions and a ceaseless search for originality, so that it is not surprising that new theories and topics are more alluring than old ones. This may help to explain why only rather general guidelines for map design have so far been established.

In cognitive studies of map reading it is impossible to study a phenomenon for which there is no externally observable evidence of mental processes resulting in a particular response. In the face of this weakness, introspection has been rediscovered in some cognitive studies. Self-reports, self-ratings and verbal protocols are now extensively used by cartographers along with more objective measures. We forget however that the utility of introspection is limited. While subjects describing the strategies they used in a map-reading task can provide useful indicators of the underlying processes, the evidence gained is of dubious reliability since many mental processes are not accessible to consciousness. It is important therefore that in any experimentation different methods and kinds of evidence are used.

Giving greater consideration to the user as an active player in the map–map reader interface also has implications for experiments. Much early experimental work was rather artificial, bearing little relationship to most map users' tasks and kinds of use. Since the mid 1970s, following the trend in psychology, cartographic experiments have had much greater ecological validity. But the advantages of ecological validity have to be traded against less stringent experiments. Greater ecological validity does allow the map user to bring knowledge, expectation and experience to bear on the experiment. But it also means that methods allowing one to research basic mental processes in controlled laboratory conditions can become difficult to handle and interpret when applied to more realistic and complex tasks (Scott 1987). Finally cartographers rarely attempt to link experiments with cartographic theory or models. This may be ascribed to the dominance of cartography as a craft, from which standpoint such considerations seem irrelevant. With greater interest in the map user as a consumer of the craft product, cartographers are beginning to relate their findings to existing theory in cognitive psychology.

Misuse of Psychological Theory

When cartographers began to borrow concepts and theories from psychology they were over-enthusiastic. Typically they ignored assumptions upon which theories were based and their applicability to the problems of map design (see, for example, Eastman and Castner (1983)). As early as 1970 Bartz had made this criticism of cartographers' research on the perception of text. While it is natural to adopt and adapt, it is often only by doing research that scientists come to know the limits of both the original and borrowed concepts.

The popularity of the Multistore Model as a theoretical framework for map reading as an information processing system is a good example of this tendency. This dominated and still dominates the psychological literature that cartographers read. Its influence in cartography has depended on its past popularity rather than its applicability. While this model in its original form has lost favour among psychologists, many cartographers retain it as their underlying theoretical framework for describing the map reading process. Although this is quite understandable in terms of its popularity, it should not prevent cartographers from looking at other models (Craik and Lockhart 1972, Baddeley 1983). A thorough grasp of the rich variety of concepts and theories in cognitive science is crucial if cartographic research is to be recognized as rigorous by other disciplines such as psychology.

Different Objectives in Cartography and Psychology

This chapter has attempted to show the advantages accruing to cartography from its relationship with psychology. Cartographic research has gradually become psychologically based. Indeed Robinson and Petchenik (1976: 3) went as far as to suggest the term Psychocartographics, 'the study of the interaction of cognition and mapping'. Cartographers and psychologists have rather different aims. Cartographers' objectives are to understand the map reader and improve the quality of map design to maximize the efficiency of communicating the desired message. Psychologists are essentially interested in the processes of communication, cognition and in individual differences. Thus, cognitive scientists have, in the main, focused on basic research, while cartographers are concerned with applied research inevitably making extensive use of the basic research as a starting point.

RESEARCH AGENDA

The Cognitive Approach and Map Use

Despite many years of research there is still no all-embracing theory of map reading. As a result empirical research and the theories proposed, appear disjointed and unconnected. Head's (1984) work on map reading as a natural language was one attempt at providing an all-embracing theory but his ideas are uncomfortable for many who disagree with the concept of cartographic language or the study of map use from a linguistic viewpoint.

We see progress coming in three ways. First, elaborating an all-embracing theory of map reading based on cognitive theory, primarily schema theory and models of map tasks (rather than linguistics) is essential. We are currently elaborating a description of map reading from a cognitive perspective, using the concept of schemata. In that, the map reader's knowledge is structured into schemata. These, in conjunction with the goals of the user, can invoke map-reading tasks and strategies by which he extracts meaning from the map. Secondly we need an inventory of the gaps in our knowledge of map reading, so that the important outstanding questions can be tackled. Thirdly the results of cognitive map use research have to be integrated with the general theory of map reading. Only then will it be possible to achieve a valid understanding of how information is used by the map reader, which is a pre-requisite for developing effective map design guidelines. To be effective such research has to be undertaken by teams.

The Cognitive Approach and Map Design

The cognitive approach is essentially aimed at understanding map use, but has little apparent relevance for map design. Research on cognitive processes and map reading may be distracting investigators from the cartographer's principal task, effective map design. Fundamental research is vital to gain an understanding of map use before one can confidently investigate its implications for map design. This must be the long-term goal, but as yet we can point to few practical applications of the cognitive approach.

In other applied fields, cognitive studies have proven crucial for establishing guidelines. But, using Human–Computer-Interaction as an example, Long and Dowell (1989) provide a strong argument as to why such an approach can never offer what it promises. Their argument is equally applicable to cartography and forms the basis of the following discussion.

In practice, cognitive research cannot contribute to the solution of a map design problem since it can only predict changes in the user's behaviour stemming from the design under a given set of conditions. It does not describe the design actions required to help the user. The predictive knowledge derived from the cognitive approach can, however, be made prescriptive. Guidelines can certainly be derived, but these can suggest only how individuals might modify their behaviour in response to a design action. They cannot assess the effectiveness of such changes. Nor can they because the prescriptive form of the guidelines has not been developed for the design of an effective map but only for the behavioural phenomena of map reading. The selection of guidelines and their application to any map would be a matter of heuristics (and so paradoxically of good practice). The application of guidelines based on the findings from cognitive research does not therefore guarantee the desired goal of an effective map.

The psychological knowledge derived from the cognitive approach cannot thus be applied directly because the knowledge is not prescriptive:

it is only explanatory and predictive. But, the use of a cognitive science approach differs in two important ways from the craft approach to map design and production. A scientific approach within a cognitive context is explicit and formal, and so supports reasoning about the derivation of guidelines and their application. Second, such knowledge would be expected to be more correct, coherent and complete than common sense knowledge about map design and map use psychology and behaviour. We are thus forced to accept a rather depressing conclusion. If one of the goals of cartographic science is to produce a general theory of map design, choosing the route of cognitive psychology will require much more effort to generate knowledge, but less to develop guidelines, but whose effectiveness can never be generalizable.

Principles of Cartographic Design

It is no surprise that cartographers making maps see little return in the way of an impressive list of design principles, since most of the effort in cognitive cartographic research has been devoted to understanding map reading processes. What is required is an approach more like that of engineering, in which the design of an effective map is solved by the specification of the design before the map is produced. (Ironically this is exactly what many well established producers of topographic map series have done for a long time.) Designing for effective performance with map design knowledge formulated as design principles can solve the map design problem through a decompositional approach. First specified at a general level of description, then systematically decomposed element by element, until a complete specification is achieved.

The objective of establishing generalizable principles of map design does not negate the role of cartography as a craft or cartography as a scientific discipline through the use of a cognitive, psychological approach. Each would have to contribute to generating design principles. For example, cartographic principles (the what) might be developed in bottom-up fashion by induction from cartographic practice (the how) and developed top-down through deductions from psycho-cartographic theory (the why).

Examples of Applications for Cognitive Cartographic Research

During the 1980s funding for cognitive cartographic research has had to compete first with automated mapping and then GIS. However, it is now recognized that further development in some areas of this new technology require a greater understanding of cognitive cartography, albeit applied.

Extending the cognitive perspective to Cartographic Expert Systems and Expert Systems for GIS

Recent discussion and research has explored the need for and the design of Cartographic Expert Systems (CES) (Robinson and Jackson 1985, Mackaness et al. 1986, Eastman 1987). A CES enables the intelligent automation of the design of maps, but efforts made to produce CES have been rather piecemeal focusing on a few aspects of map design. None of them can be considered as holistic or complete CES. Mackaness and Scott (1988) have pointed out that map design is a multi-domained task and that the apparent sequential order of the design process is deceptive since the cartographer will seemingly take up tasks and strategies at random. A fundamental drawback is the absence of in-depth knowledge of how cartographers set about designing a map. Remedying this will require the use of knowledge elicitation and protocol analysis. Cognitive studies of map makers, as opposed to map users, can form a major part of this knowledge elicitation process. In this way we may begin to understand the tasks and strategies, processes and knowledge used to produce maps. Indeed Robinson and Petchenik (1976) argued that emphasis in cartographic research should shift to the cognition and perceptual activities of not only map readers but also the map makers.

As it is now easy for a 'non-cartographer' to produce maps as a by-product or end-product of GIS, there is cause for concern over the quality of displays. It would be logical to develop cartographic expert systems for GIS which would suit the non-cartographer. Apart from needing to know how cartographers design maps, it will be necessary to design, from a cognitive standpoint, an advisor system usable by non-experts and expressed in simple terms (Muller, Chapter 1).

HCI issues of Automated Mapping Systems and GIS

There is a growing awareness that the design of the user interface in Automated Mapping Systems (AMS) and GIS is crucial for their efficiency, effectiveness and acceptability (Medyckyj-Scott et al. 1990). Unfortunately such interfaces are often difficult to use, involve over-complex forms of user interaction and do not suit the user's working environment. The majority of user interaction with such systems is through maps displayed on screens. It is not known whether the considerable knowledge derived from cognitive cartographic studies (on paper maps) is directly applicable to VDUs. The ability for the user to zoom and pan on such displays, to add and subtract map detail, to make graphics-based queries and to produce dynamic maps are all tasks not normally performed by the user in a traditional map use context. Cognitive cartographic research, if applied to this issue, ought to improve the user interfaces of future systems.

In searching for ways to improve the effectiveness of GIS, considerable effort is also being expended in an attempt to discover how people think about space and express this knowledge; the concepts used to describe the relations between objects in space and the importance of maps for understanding space (Peuquet 1988, Mark and Frank 1989). Cognitive cartography has a crucial role to play in this emerging area.

In-car navigation systems

Since the early 1980s there has been considerable interest in the development of computer based in-car, or Vehicle Navigation Assistance systems (VNAs). Many, although not all these systems, include some form of map display (Claussen and Mark, Chapter 8). Petchenik (1989) has argued that the current maps used in such systems have a number of shortcomings. Moreover, the mental-task load of a driver during a journey cannot accommodate the additional tasks of map-based navigation. Attempts to overcome these problems have led to the newer VNA systems relying more on spoken instructions to the driver. These unfortunately also have their drawbacks. Evidently both visual and audible cues appear desirable, but a more thorough knowledge of cognitive processes and tasks used by drivers and what role maps play in navigation are essential for improving VNA systems. Recent research includes that on individual differences in way finding (Gould 1989) and the cognitive processes and spatial knowledge required to navigate (Mark and McGranaghan 1986, Freundschuh 1989) but the cartographers contribution to this area is currently limited. Studies by Pailhous (1970) and Chase (1983) on how taxi-drivers navigate in cities deserve close scrutiny.

Mapwork in education

There is no lack of advice in textbooks and in prescriptions handed out by expert working parties (eg DES 1990: 84) on the teaching of map reading. Despite the rather comforting conclusions reached in a recent review by Ottoson (1988), there remains a strong suspicion that many practising school teachers know what methods work, even if they do not know why. In addition, surprisingly all too little attention appears to have been paid to atlas design and use in schools. However, a recent volume of essays (Carswell et al. 1987) has shown how productive it is to bring together experts from different disciplines and that the cartographer needs to learn from others if atlases are to serve their purpose of stimulating children (and adults) to learn more about their world. The cognitive approach has much to offer to the teaching of map reading by explaining why it is that some teaching succeeds better than others.

PRACTITIONER'S SUMMARY

There are no obvious achievements to which we can point. However the fairly widespread awareness of the cognitive approach, coupled with evidently unproductive approaches through communication theory, linguistic theory and psychophysics, have focused attention on the mental set and behaviour of the map reader. But this has thrown into relief the need to understand the procedures followed by the map designer. Knowledge of these kinds is essential for the development of useful and valid cartographic expert systems. It is now widely accepted that findings from research on conventional paper maps can not be automatically applied to displays on VDUs. Currently our knowledge of map use in automated mapping and GIS is very limited, but the cognitive approaches employed for studying the use of paper maps can be adapted to this technological environment.

PROJECTIVE SUMMARY

1. Elaborate an all-embracing theory of map reading based on cognitive theory.
2. Make an inventory of gaps in knowledge of map reading.
3. Integrate cognitive research on map use with the general theory of map reading.
4. Develop guidelines for map design deductively from cognitive–psychological theory to match the development of guidelines for map design derived inductively.
5. Extend the cognitive perspective to holistic Cartographic Expert Systems.
6. Design a cognitively based advisor system to help non-experts make maps.
7. Understand, through cognitive processes and tasks, how drivers navigate, so that Vehicle Navigation Assistance systems can be made more effective.
8. Apply the cognitive approach to teaching map-reading skills.

REFERENCES AND BIBLIOGRAPHY

Anderson J R (1978) Arguments concerning representations for mental imagery. *Psychological Review* **85**: 249–77

Anderson J R (1980) *Cognitive psychology and its implications*. San Francisco, Freeman

Aslanikashvili A F (1968) *Cartography, problems of general theory*. Tbilisi, Metsniereba (in Georgian) 298pp.

Aslanikashvili A F (1974) *Metacartography, basic problems*. Tbilisi, Metsniereba (in Russian) 125pp.

Atkinson R C, Shiffrin R M (1968) Human memory : A proposed system and its control processes. In: Spence K W and Spence J T (ed.) *The psychology of learning and motivation: Advances in research and theory* (*Vol.*2). New York, Academic Press

Baddeley A D (1983) Working memory. *Philosophical Transactions of the Royal Society of London* B **302**: 311–24

Baddeley A D (1985) Domains of recollection. In: Aitkenhead A M and Slack J M (ed.) *Issues in cognitive modeling*. London, Lawrence Erlbaum Associates: 209–28

Bartz B (1970) The experimental use of the search task in an analysis of type legibility in cartography. *The Cartographic J.* **7** (2): 147–67

Bartz B (1969) Search : An approach to cartographic type legibility measurement. *The Journal of Typographic Research* **3** (4): 387–9

Bertin J (1967) *Semiologie graphique*. Paris, Gauthier-Villars

Blades M, Spencer C (1986*a*) Young children's strategies when using maps with landmarks. Department of Psychology, University of Sheffield, UK

Blades M, Spencer C (1986*b*) The implications of psychological theory and methodology for cognitive cartography. *Cartographica* **23** (4): 1–13

Blades M, Spencer C (1987*a*) How do people use maps to navigate through the world? *Cartographica* **24** (3): 64–75

Blades M, Spencer C (1987*b*) Young children's strategies when using maps with landmarks. *Journal of Environmental Psychology* **7**: 201–17

Blaut, J.M. (1987) Notes towards a theory of mapping behavior. *Children's Environments Quarterly* **4** (4): 27–34

Bluestein N, Acredolo L (1979) Developmental changes in map-reading skills. *Child Development* **50**: 691–7

Board C (1967) Maps as models. In: Chorley R J and Haggett P (ed.) *Models in Geography*. London, Methuen: 671–725

Board C (1977) The geographers contribution to evaluating maps as vehicles for communicating information. *International Yearbook of Cartography* **17**: 47–59

Board C (1978) Map reading tasks appropriate in experimental studies in cartographic communication. *The Canadian Cartographer* **15** (1): 1–12

Board C (1981) Cartographic communication. In: Guelke L (ed.) Maps in Modern Geography. *Cartographica* **18** (2): 42–78

Board C (1984) Higher order map-using tasks: Geographical lessons in danger of being forgotten. *Cartographica*, Monograph 31, **21** (1): 85–97

Carswell R J B, de Leeuw G J S, Waters N M (ed.) (1987) Atlases for Schools. Design principles and curriculum perspectives. *Cartographica*, Monograph 36, **24** (1)

Castner H W (1983) Research questions and cartographic design. In: Taylor D R F (ed.) *Graphic communication and design in contemporary cartography*. Chichester, Wiley and Sons Ltd: 87–114

Castner H W, Lywood D W (1978) Eye movement recording: Some approaches to the study of perception. *The Canadian Cartographer* **15** (2): 142–50

Cerney J W, Wilson B (1976) The effect of orientation on the recognition of simple maps. *The Canadian Cartographer* **13** (2): 132–8

Chang K (1977) Visual estimation of graduated circles. *The Canadian Cartographer* **14** (2): 130–8

Chang K (1980) Circle size judgements and map design. *The American Cartographer* **7** (2): 155–62

Chang K, Antes J, Lenzen T (1985) The effect of experience on reading topographic relief information: analysis of performance and eye movements. *The Cartographic J.* **22** (2): 88–94

Chase W G (1983) Spatial representations of taxi drivers. In: Rogers, D. and Sloboda J A (ed.) *The Acquisition of Symbolic Skills*. NATO Conference Series 3, Human Factors. New York, Plenum Press: 391–405

Chase W G, Simon H A (1973) The mind's eye in chess. In: Chase W G (ed.) *Visual information processing*. New York, Academic Press

Cox C W (1976) Anchor effects and the estimation of graduated circles and squares. *The American Cartographer* **3** (1): 65–74

Craik F I M, Lockhart R S (1972) Levels of processing: a framework for memory research. *Journal of Verbal Learning and Verbal Behaviour* **11**: 671–83

Cromie B W (1977) Contour design and the topographic map user. *The Canadian Surveyor* **31** (1): 34–40

Cross K D, Rugge S M, Thorndyke P W (1982) Cognitive processes in interpreting the contour-line portrayal of terrain relief. Personnel and Training Research Programs, Psychological Sciences Division, Office of Naval Research, Contract No. NOOO14–80 – C – 0508, Anacapa Sciences Inc., Santa Barbara, California

Cuff D J (1973) Colour on temperature maps. *The Cartographic Journal* **10** (1): 17–20

DES (Department of Education and Science) (1990) Geography for ages 5 to 16. Proposals of the Secretary of State for England and Wales. London, National Curriculum Council

De Lucia A A (1976) How people read maps. American Congress of Surveying and Mapping Technical Papers, Spring Convention: 135–44

Dent B D (1972) Visual organisation and thematic map communication. *Annals, Association of American Geographers* **62**: 79–93

Dobson M W (1977) Eye-movement parameters and map reading. *The American Cartographer* **4**: 39–58

Dobson M W (1979) Visual information processing during cartographic communication. *The Cartographic Journal* **16** (1): 14–20

Eastman J R (1977) Map complexity: An information approach. Unpublished Master's thesis, Queens University, Kingston, Ontario.

Eastman J R (1982) Graphic organisation and the structuring of map data into memory. Unpublished PhD dissertation, Boston University, USA

Eastman J R (1985*a*) Cognitive models and cartographic design research. *The Cartographic Journal* **22** (2): 95–101

Eastman J R (1985*b*) Graphic organisation and memory structures for map learning. *Cartographica* **22** (1): 1–20

Eastman J R, Castner H W (1983) The meaning of experience in task-specific map reading. In: Taylor D R F (ed.) *Graphic communication and design in contemporary cartography*. Chichester, Wiley and Sons Ltd: 115–48

Eastman R (1987) Graphic syntax and expert systems for map design. *Proceedings, ASPRS–ACSM Annual Convention*, Baltimore, Maryland **4**: 87–96

Eden C (1988) Cognitive Mapping. *European Journal of Operational Research* **36** (1): 1–13

Edwards A L (1970) *Experimental Design in Psychological Research*, (3rd Edition). London, Holt, Rhinehart and Winston

Ekman G, Lindman R, William-Olsson W (1961) A psychophysical study of cartographic symbols. *Perceptual and Motor Skills* **13**: 355–68

Eley M G (1981) Imagery processing in the verification of topographical cross-sections. *Educational Psychology* **1**: 39–48

Eley M G (1983*a*) Representing the cross-sectional shapes of contour mapped landforms. *Human Learning* **2**: 279–94

Eley M G (1983*b*) Judging relative proximity in a geographical region. *Human Learning* **2**: 127–39

Fechner G T (1860) *Elemente der Psychophysik*. Translated by Alder H E (1966) Elements of psychophysics. New York, Holt, Rhinehart and Winston

Federer W T (1955) *Experimental design*. New York, Macmillan

Fergusson G A (1966) On transfer and human ability. *Canadian Journal of Psychology* **10**: 121–30

Flannery J J (1956) The graduated circle: A description, analysis and evaluation of a quantitative map symbol. Unpublished PhD thesis, University of Wisconsin, Madison, Wisconsin, USA

Freitag U (1971) *Semiotik und Kartographie, über die Anwendung kybernetischer Disziplinen in der theoretischen Kartographie*. Kartografische Nachrichten, no.5: 171–82

Freundschuh S M (1987) Assessing the abilities of pre-literate children to use maps. Unpublished Master's thesis, Department of Geography, State University of New York at Buffalo

Freundschuh S M (1989) Does 'Anybody' really want (or need) vehicle navigation aids. *First IEEE Vehicle Navigation and Information Systems Conference*, Toronto, Ontario: 439–42

Freundschuh S, Gould M D, Mark D M, Egenhofer M J, Kuhn W, McGranaghan M, Soteria S (1989) *Working bibliography on 'Langauges of Spatial Relations'*. National Center for Geographic Information and Analysis, USA, Technical Paper 89–10

Gamezo M U, Rubakhin V F (1964) The role of spatial concepts in map reading and the interpretation of aerial photographs. In: Ananyev B G and Lomov B F (ed.) *Problems of spatial perception and spatial concepts*. Washington DC, National Aeronautics and Space Administration, Technical Translation No. F – 164

Gatrell A C (1974) *On the complexity of maps*. Pennsylvania State University Papers in Geography No.11

Gilhooly K J, Wood M, Kinnear P R, Green C (1988) Skill in map reading and memory for maps. *The Quarterly Journal of Experimental Psychology* **40 A** (1): 87–107

Gilmartin P P (1981) The interface of cognitive and psychophysical research in cartography. *Cartographica* **18** (3): 9–20

Gilmartin P P, Patton J C (1984) Comparing the sexes on spatial abilities: Map-use skills. *Annals, Association of American Geographers* **74** (4): 605–19

Gould D M (1989) Considering individual cognitive ability in the provision of usable navigation assistance. *First IEEE Vehicle Navigation and Information Systems Conference*, Toronto, Ontario: 443–7

Greeno J G, Simon H A (1988) Problem solving and reasoning. In: Atkinson R C, Hernstein H, Lindzey G and Luce R D (ed.) *Steven's handbook of experimental psychology*. New York, Wiley

Griffin T L C (1983) Problem solving on maps — the importance of user strategies. *The Cartographic Journal* **20** (2): 101–9

Guelke L (1979) Perception, meaning and cartographic design. *The Canadian Cartographer* **16** (1): 61–79

Hackman J R (1969) Toward understanding the role of tasks in behavioural research. *Acta Psychologica* **31**: 97–128

Head G C (1972) Land–water differentiation in black and white cartography. *The Canadian Cartographer* **9** (1): 25–38

Head G C (1984) The map as natural language: A paradigm for understanding. *Cartographica*, Monograph 31, **21** (1): 1–32

Herskovits A (1987) *Spatial propositions in English*. Cambridge, Massachusetts, Cambridge University Press

Hopkin V D, Taylor R (1979) *Human factors in the design and evaluation of aviation maps*. Neuilly-sur-Seine, NATO AGARDograph No. AG225

Jenks G F (1975) The evaluation and prediction of visual clustering in maps symbolised with proportional circles. In: Davis J C and McCullagh M J (ed.) *Display and analysis of spatial data*. London, John Wiley: 311–27

Jenks G F, Caspall F C (1967) Vertical exaggeration in three dimensional mapping. United States Government, Office of Naval Research (Geography Branch), Project No. NR 38–146, Technical Report No.2, Contract No. NONR 583(15)

Jenks G F, Caspall F C (1971) Error on Choropleth maps: definition, measurement, reduction. *Annals Assoc. American Geographers* **61** (2): 217–44

Keates J S (1982) *Understanding maps*. London, Longman

Kerst S M, Howard J H (1978) Memory psychophysics for visual area and length. *Memory and Cognition* **6**: 327–35

Kimerling A J (1975) A psychophysical study of equal value grey scale. *Proceedings of the American Congress of Surveying and Mapping*, Washington, USA: 234–47

Kosslyn S M, Pinker S, Smith G E, Shwartz S P (1979) On the demystification of mental imagery. *The Behavioural and Brain Sciences* **2**: 532–81

Kuipers B (1978) Cognitive modelling of the map user. In: Dutton G (ed.) *Proceedings, First International Study Symposium on Topological Data Structures for Geographic Information Systems*, Vol. 7, Spatial Semantics: Understanding and Interacting with Map Data: Kuipers/1–Kuipers/11

Lachman R, Lachman J L, Butterfield E C (1979) *Cognitive psychology and information processing: An introduction*. Hillsdale, NJ, Lawrence Erlbaum Associates

Liben L S, Downs R M (1989) Understanding maps as symbols: The development of map concepts in children. In: Reese H W (ed.) *Advances in child development and behavior*, Vol 22. San Diego, California, Academic Press

Lloyd R (1982) A look at images. *Annals, Association of American Geographers* **72** (4): 532–48

Lloyd R, Steinke T (1984) Recognition of disorientated maps: The cognitive process. *The Cartographic Journal* **21**: 55–9

Lloyd R, Steinke T (1985*a*) Comparison of quantitative point symbols / The cognitive process. *Cartographica* **22** (1): 59–77

Lloyd R, Steinke T (1985*b*) Comparison of qualitative point symbols / The cognitive process. *The American Cartographer* **12**: 156–68

Long J, Dowell J (1989) Conceptions of the discipline of HCI: Craft, Applied Science, and Engineering. In: Sutcliffe A and Macaulay L (ed.) *People and computers V*. Cambridge, Cambridge University Press: 9–32

Mackaness W A, Scott D J (1988) The problems of operationally defining the map design process for cartographic expert systems. *Proceedings, Austra Carto III*, Aug.22–26th, 7th Australian Cartographic Conference, Sydney, Australia: 715–23

Mackaness W A, Fisher P F, Wilkinson G (1986) Towards a cartographic expert system. *Proceedings, Auto Carto London*, 14–19 Sept. **1**: 578–87

Mark D M, Frank A U (1989) Concepts of space and spatial relations. *Proceedings, Ninth International Symposium on Computer Assisted Cartography*, Baltimore, Maryland: 538–56

Mark D M, McGranaghan M (1986) Effective provision of navigation assistance to drivers: A cognitive science approach. *Proceedings, Auto Carto London*, 14–19 Sept.: 3309–408

McCleary G F (1970) Beyond simple psychophysics : Approaches to the understanding of map perception. *Proceedings of the American Congress of Surveying and Mapping*, Technical Papers, 30th Annual Meeting, March 1970: 189–209

McGuigan F J (1978) *Experimental Psychology: A Methodological Approach*. Englewood Cliffs, New Jersey, Prentice-Hall Inc

McNamara T P (1986) Mental representations of spatial relations. *Cognitive Psychology* **18**: 87–121

Medyckyj-Scott D J, Newman I A, Ruggles C, Walker D (1990) Usability testing in user interface design for GIS and the role of rapid prototyping. *Proceedings, First European GIS Conference*, April 10–13, Amsterdam, Netherlands: 737–46

Meihoeffer H J (1969) The utility of the circle as an effective cartographic symbol. *The Canadian Cartographer* **9** (2): 105–17

Minsky M (1975) A framework for representing knowledge. In: Winston P H (ed.) *The psychology of computer vision*. New York, McGraw-Hill

Monkhouse F J, Wilkinson H R (1952) *Maps and diagrams: their compilation and construction*. London, Methuen

Morrison J L (1974) A theoretical framework for cartographic generalisation with emphasis on the process of symbolisation. *The International Yearbook of Cartography* **14**: 115–27

Morrison J L (1975) The science of cartography and its essential processes. Paper presented at the *First International Symposium on Cartographic Communication*, London, 15–19 Sept.

Morrison J L (1976) The relevance of some psychophysical research to simple map reading. Paper presented at 8*th International Cartographic Conference*, Moscow

Moyer R, Bradley D, Sorensen M, Whiting J, Mansfield D (1978) Psychophysical functions for perceived and remembered size. *Science* **200**: 330–2

Muehrcke, P C (1973) Visual pattern comparison in map reading. *Proceedings of the Association of American Cartographers* **5**: 193–4

Neisser U (1967) *Cognitive psychology*. New York, Appleton-Century-Crofts

Neisser U (1976) *Cognition and reality*. San Francisco, WH Freeman and Co

Newell A, Simon H (1972) *Human problem solving*. Englewood Cliffs, New Jersey, Prentice-Hall

Olson J M (1975) Autocorrelation and visual map complexity. *Annals, Association of American Geographers* **65** (2): 189–204

Olson J M (1976) A co-ordinated approach to map communication improvement. *The American Cartographer* **3** (2): 151–9

Olson J M (1979) Cognitive cartographic experimentation. *The Canadian Cartographer* **16** (1): 34–44

Ottoson T (1988) What does it take to read a map? *Cartographica* **25** (4): 28–35

Pailhous J (1970) *La représentation de l'espace urbain. L'exemple du chauffeur du taxi*. Paris, Presses Universitaires de France

Paivio A (1969) Mental Imagery in associative learning and memory. *Psychological Review* **76**: 241–63

Palmer S E (1975) Visual perception and world knowledge: Notes on a model of sensory–cognitive interaction. In: Norman D A and Rumelhart D E (ed.) *Explorations in cognition*. San Francisco, Freeman

Petchenik B B (1975) Cognition in cartography. *Proceedings of the Symposium on Computer Assisted Cartography*: 183–93

Petchenik B B (1983) A map-maker's perspective on map design research 1950–1980. In: Taylor D R F (ed.) *Graphic communication and design in contemporary cartography*. Chichester, Wiley and Sons Ltd: 37–68

Petchenik B B (1989) The nature of navigation: Some difficult cognitive issues in automatic vehicle navigation. *First IEEE Vehicle Navigation and Information Systems Conference*, Toronto, Ontario: 43–8

Peterson M P (1978) *An evaluation of no-class choropleth mapping*. Unpublished Masters thesis, Boston University, USA

Peterson M P (1985) Evaluating a maps image. *The American Cartographer* **12** (1): 44–55

Peuquet D J (1988) Representations of geographic space: Towards a conceptual synthesis. *Annals, Association of American Geographers* **78**: 375–94

Phillips R J, De Lucia A, Skelton N (1975) Some objective tests of the legibility of relief maps. *The Cartographic Journal* **12** (1): 39–46

Phillips R J, Noyes E, Audley R (1978) Searching for names on maps. *The Cartographic J* **15** (1): 72–7

Phillips W A (1983) Short-term visual memory. *Philosophical Transactions of the Royal Society of London B* **302**: 295–309

Pick H L, Acredolo L P (ed.) (1983) *Spatial orientation: Theory, research and application*. New York, Plenum Press

Pinker S (ed.) (1985) *Visual cognition*. Cambridge, Massachusetts, MIT Press

Presson C C (1982) The development of map reading skills. *Child Development* **53**: 196–9

Prince V S (1974) Characteristics of map reader recall of a single regional shape. MA thesis, University of Kansas, USA

Raisz E (1938) *Principles of cartography* (1st edition). New York, McGraw-Hill (2nd edition, 1948)

Robinson A H (1952) *The look of maps: An examination of cartographic design*. Madison, USA, University of Wisconsin Press

Robinson A H (1967) Psychological aspects of colour in cartography. *The International Yearbook of Cartography* **7**: 50–9

Robinson A H, Petchenik B B (1976) *The nature of maps: Essays towards understanding maps and mapping*. London, University of Chicago Press

Robinson G J, Jackson M J (1985) Expert systems in map design. *Proceedings, Auto Carto 7 Digital Representation of Spatial Knowledge*, Washington DC, March 11–14: 430–9

Rock I (1974) The perception of disorientated figures. *Scientific American* **230**: 78–85

Rumelhart D E, Norman D A (1985) Representation of knowledge. In: Aitkinhead A M and Slack J M (ed.) *Issues in cognitive modelling*. London, Erlbaum: 15–62

Rumelhart D E, Ortony A (1977) The representation of knowledge in memory. In: Anderson R C, Spiro R J and Montague W E (ed.) *Schooling and the acquisition of knowledge*. Hillsdale, NJ, Lawrence Erlbaum Associates

Schank R C, Abelson R (1977) *Scripts, plans, goals and understanding*. Hillsdale, NJ, Lawrence Erlbaum Associates

Scott D J (1987) Mental imagery and visualisation: Their role in map use. Unpublished PhD thesis, University of London

Shaw M E (1963) Scaling group tasks: A method for dimensional analysis. Technical Report No.1, University of Florida

Shepard R N, Chipman S (1970) Second order isomorphisms of internal representations: The shape of states. *Cognitive Psychology* **1**: 1–17

Shepard R N, Hurwitz S (1985) Upward direction, mental rotation, and discrimination of left and right turns in maps. In: Pinker S (ed.) *Visual cognition*. Cambridge, Massachusetts, MIT Press: 161–94

Shimron J (1978) Learning positional information from maps. *The American Cartographer* **5** (1): 9–19

Sholl M J, Egeth H E (1982) Cognitive correlates of map-reading ability. *Intelligence* **6**: 215–30

Simitus Z M, Barsam H F (1983) Terrain visualisation and map reading. In: Pick H L and Acredolo L P (ed.) *Spatial orientation: Theory, research and application*. New York, Plenum Press

Simon H A (1985) Information processing theory of human problem solving. In: Aitkinhead A M and Slack J M (ed.) *Issues in cognitive modeling*. Hillsdale, NJ, Lawrence Erlbaum Associates

Sophian C (1984) *Development of cognitive skills*. Hillsdale, NJ, Erlbaum

Steinke T, Lloyd R (1983) Images of maps: A rotation experiment. *Professional Geographer* **35** (4): 455–61

Stevens S S (1975) *Psychophysics: Introduction to its perceptual, neural and social prospects*. New York, Wiley Inter-science

Thorndyke P W, Hayes-Roth B (1982) Differences in spatial knowledge acquired from maps and navigation. *Cognitive Psychology* **14**: 560–89

Thorndyke P W, Stasz C (1980) Individual differences in procedures for knowledge acquisition from maps. *Cognitive Psychology* **12**: 137–75

Ullman S (1985) Visual routines. In: Pinker S (ed.) *Visual cognition*. Cambridge, Massachusetts, MIT Press: 97–160

Uttal D H, Wellman H M (1989) Young childrens representation of spatial information acquired from maps. *Developmental Psychology* **25**: 128–38

Waugh N C, Norman D A (1965) Primary memory. *Psychological Review* **72**: 89–104

Wheate R D (1978) Commensurability v. imageability: A re-assessment of the role played by shaded relief on topographic maps. Unpublished M.A. thesis, Queens University, Kingston, Canada

Williams R L (1956) Statistical symbols for maps: Their design and relative values. Yale University Map Laboratory Report, New Haven, Conn., USA

Williamson J (1987) Schema development in geography pupils' understanding of maps. Unpublished PhD thesis, Queens University, Belfast, Northern Ireland

Wood M (1968) Visual perception and map design. *The Cartographic Journal* **5** (1): 54–64

Wood M (1972) Human factors in cartographic communication. *The Cartographic Journal* **9** (2): 123–32

Wood M, Kinnear P (1984) Chunking and pattern recognition in contour map reading and learning situations. *Proceedings of ICA 12th International Conference*, Technical papers No.1: 666–7

Wood R E (1986) Task complexity: Definition of the construct. *Organizational Behaviour and Human Decision* **37**: 60–82

Chapter 11

INSTITUTIONAL AND SOCIETAL COMPONENTS OF CARTOGRAPHIC RESEARCH

Nicholas R. Chrisman
Geography DP – 10, University of Washington, Washington 98195, USA

Just as technological advances are changing the ways cartographers make maps, there is a parallel impact on the institutions for which they work. For many years, cartographic research could be characterized as 'tool development' with an implicit belief in the inevitability and desirability of 'progress'. Cartographic research should not assume that all new tools will be evaluated on some objective scale of efficiency and effectiveness. Despite the best of technology, some tools will succeed or fail based on their institutional ramifications and even on their connections to greater societal goals. This chapter considers the breadth of concerns for institutional constraints and societal outcomes.

BACKGROUND: RESEARCH IN CARTOGRAPHY

In the 1990s, research in cartography continues the recent emphasis on development of tools for the continued automation of cartography. The previous chapters of this book were written by international experts and emphasize an abstract view of tools with universal application. This chapter takes a different position. While accepting the validity of the preceding chapters as research efforts, it is necessary to add a cautionary note from the perspectives of institutional and social context.

The starting point of this chapter is not an abstract view of cartographic information. Instead, cartographic research should understand who uses cartographic information and why. Most cartography is not carried out in an academic laboratory for individual purposes. Cartographers work inside institutions, some private but many public, and the nature of these institutions influences the current reality as well as the future of cartography. In turn, these institutions serve, to some degree or another, the general goals of the particular society in which they are embedded. An ardent cartographic technologist might accept that there are institutional and social aspects of cartography, but might fail to appreciate that these concerns are a part of the most pressing research agenda. This chapter attempts to make the argument that research on tool development must be balanced with research on the outcomes in institutions and society. For, our tools are only truly effective if they are actually used.

NEITHER ART NOR SCIENCE

The commonly accepted definition of cartography begins with the assertion that cartography is at once an art and a science (ICA 1973). This chapter does not seek to redefine cartography, but it must take note of some divergences between the accepted view of cartography and cartography as it is actually practised.

In purest form, an artist is free to express a particular vision or message often by testing the bounds of the chosen medium. Cartographers have long accepted that the constraints of mapping make it hard to act as freely as an artist would expect. Of course, there is some ingredient of art in cartography — an ingredient which must be cherished and nurtured.

The research agenda discussed in the previous chapters provides very little coverage (except perhaps for Chapter 10) of the artistic component of cartography because it emphasizes the scientific component. Proclaiming the scientific in cartography may be an unannounced purpose of publication of a volume on the research agenda at a General Assembly of the International Cartographic Association. Certainly, there have been many advances in cartographic method since the creation of the ICA. Yet, for all of the advances of the computer age, most cartographers do not practise in a purely scientific context. Scientific cartography attempts to study alternative tools, measured on objective grounds of efficiency and effectiveness. To a large extent, these goals are desirable, as long as the measures are appropriate. Scientific cartography fails to be a model for practise first, if the objective measures are not so easily established, and second, if the constraints of societies and institutions are ignored. To the cartographer working in a municipal information system or a national topographic agency, the procedures used may have some connection to scientific research, but they also will be firmly connected to the budget of the organization and its management structure.

Rather than either an art or a science, the practice of cartography should be considered as a form of regulated utility. The distinction is that practising cartographers do not have very much control over the content of their maps and databases. Each cartographic organization operates under a set of *mandates* which establish the objectives of the organization (Chrisman and Niemann 1985). In many cases, mandates are specified in statute, decree, ordinance, administrative code or some combination of such legal instruments. In cases where cartographic procedures are not clearly established through an external legal process, they may be specified through a set of instructions passed on inside the organization. In any case, it is important to recognize that cartographers frequently operate under a form of public trust.

As with other regulated utilities, the public is not particularly interested in the details of the process. The final product is what the public sees. The experts and technocrats tend to have control over the details, as long as the system seems to function. Thus, much of the research on scientific cartography has concentrated on re-engineering the procedures, assuming that the products are given. This approach certainly applied to the spaghetti era of automated drafting, but it also applies to more recent research. While the products produced under previous technology may have been perfectly adequate, considering the levels of resources available, modern technology may be able to deliver a different product which better serves the ultimate goals. This is the true challenge to cartographers in a technological revolution: to deliver new tools which open up new ways to manage spatial information. This can only be achieved by studying the social and institutional context of cartographic information handling to understand the root causes and ultimate needs.

TECHNOLOGICAL CHANGE AND CARTOGRAPHY

Cartography has advanced through a dramatic series of technical innovations during this century, so technological transitions are nothing new. In large part, research in the history of cartography involves reconstruction of the technology of an earlier period without a comparative intent. Recent research in the history of cartography (Blakemore and Harley 1980, Harley and Woodward 1987, Harley 1989) shows a substantial broadening of interest and a focus on the meaning of cartography, not just the technology. Monmonier (1985) entitled a book *Technological Transition in Cartography* and *The American Cartographer* (Tomlinson and Petchenik 1988) devoted an issue to personal recollections of the 'revolution', but much more remains to be done. We have to get beyond the classic first sentence which has adorned the cartographic literature for too many years. Perhaps the most pungent form of this sentence was the first sentence of the POLYVRT manual (Laboratory for

Computer Graphics, 1973: 1): 'Recent years have witnessed an upsurge ...' Yes, of course there has been an increase in the use of computers, but why and to what avail?

One direction for research is to consider the economic and social consequences of the shifts in cartographic technology. Monmonier (1985: 14) made this a key point in his treatment of cartographic technology; 'If the fifteenth century saw a revolution in publishing and the nineteenth saw a revolution in design, the late twentieth and early twenty-first centuries must see a revolution in organization and management.'

But what kinds of changes can we expect? There are fundamental differences between most of the technological changes which occurred in the first part of this century and the recent wave of computer systems. Many previous innovations, such as airphotography, had the effect of centralizing the production of maps and promoting efficiency through standardized production in volume. Such technical concentrations are not neutral, they have led to the large national mapping agencies around the world and a concentration of disciplinary skills. The new computerized technology turns the tide from centralized, standardized production towards an emphasis on flexibility. A flexible industry can respond to the actual demands more specifically, differentiating products instead of providing a standardized one-size-fits-all. This development, too, is not neutral. Existing institutions, with their centralist heritage and missions may not appreciate or exploit the full possibilities offered by the technology. A review of recent history can expand on these characterizations.

CENTRALIZING TENDENCIES: ECONOMIES OF SCALE

The twentieth century has seen a number of technological changes in cartography long before the computer. Acceleration in the rate of change was even noticed over fifty years ago (Raisz 1937). Most of the innovations of the early twentieth century, which included airphotography, photographic duplication, offset printing and other technologies, involved increased productivity through capital investment. The quality of the product could be increased through the new machines, but the final distribution was still in the form of a printed map. In order to gain the full value, this printed map had to serve many map readers. Thus there was a push for standardized products and national conformity to standards. The national topographic surveys (USGS, OS, IGN, etc) are the institutional consequence of this drive for centralized, standard products. To a great extent, such centralized organizations make it easier to develop and implement new technologies, because the new skills need only be concentrated where actually needed. In addition to the centralization, these organizations involved division of labour. Indeed, these organizations have all the characteristics described as the 'Fordist' model by economic geographers (Scott and Storper 1986). Whatever the terminology, it is important to recognize the linkage between expensive capital equipment, scarce training and uniform, standardized products.

DECENTRALIZING TENDENCIES: FLEXIBLE PRODUCTION

The computer was treated originally as a similar technology to airphotography. It certainly was difficult to operate and required significant investment just to get it started. The large central organizations were among the few pioneers in automated cartography (see for instance the Proceedings of the International Symposium on Map and Chart Digitizing in 1969). There were thoughts of centralized databanks of national (as in CGIS (Tomlinson 1974)) or at least state scope in the USA.

However, the computer age turned out to be quite different. The idea of centralized databanks was attempted in the USA during the 1970s. Most of the statewide efforts failed (Mead 1981), as much for institutional reasons as technological ones. The technology diffused from its origins to an ever expanding set of users. In the language of the 1980s, the computer was 'empowering' or equalizing. The software encapsulated expert knowledge about map making and map analysis and replicated that knowledge. In the place of the division of labour, the computer age implied a return to the artisanal approach to map making where one person could take a project through all steps from data collection to final production (Bie 1984). In addition, the product need not be a standardized printed map sheet. Each query can create its own graphic, particularly tailored to that message. In short, the rules of the game are altered by the computer, though the players in the game may not catch on as quickly. The computer revolution fits fairly closely to the movement in economic organization known as 'flexible production', seen as a contrast to 'Fordist' organization. Flexible production may substitute small, specialized artisanal firms for the industrial dinosaurs of the assembly line. Of course, both centralization and flexibility are old forms of organization which reappear from time to time in connection with particular technologies.

HISTORICAL SEQUENCE OF TECHNOLOGICAL CHANGE

From a quick review of recent events, each technology has its particular impact on the people and institutions who implement the technology. In addition, the historical juxtaposition of technologies has an effect as well.

The computer tendency towards flexible production occurs after a period of centralization, so the effect will be different. The discipline of cartography, and particularly the allied disciplines such as photogrammetry, may not recognize the shift in direction.

In third world countries, there are additional twists to the story. Technology does not always recapitulate the phylogeny or sequence which occurred in Europe and North America. Technology tends to arrive for reasons external to the implementing country.

'Such systems, having relatively high capital cost, are invariably a key item in some aid package. Normally, various donor agencies negotiate with separate arms of Government, with the donors committed in advance to 'selling' specific hardware and software products from their own nationals, along with the expertise to operate them. This needless proliferation of systems, none fully maximized, inevitably results in waste and inefficiency'. (Eyre 1989: 367)

Eyre's experience in Jamaica represents experiences in many other countries. Many of the difficulties in transferring GIS technology have institutional roots. Some evoke echoes of colonialist zones of influence within a country's bureaucracy.

RESEARCH DIRECTIONS

The purpose of this chapter is to discuss the goals of cartography and future research directions. As such, it must speculate rather than presenting specific backing for its assertions. Whatever proof or support will have to arise over time.

The primary plea of this chapter is to balance the interest, evidenced in the rest of this volume, in tool development with the interest in the outcomes of using those tools. The study of outcomes is the study of what happens when a tool is used or not used; it is the study of what societal goals are better fulfilled and which segments of society gain or lose. To conduct this research, cartographers must place their efforts firmly in context of social, economic, political and cultural circumstance.

This plea should not be taken as an attack on scientific research in cartography, but rather as an attempt to conduct that research more effectively. There certainly are components of cartography, such as the axiomatic structure of geometry, which will apply universally. However, as demonstrated in the case of airphotography, technology is not neutral. Cartographers must take responsibility for the directions we take, and attempt to conduct research which serves our ultimate goals. This section reviews two specific tracks for future research in the field of social and institutional context.

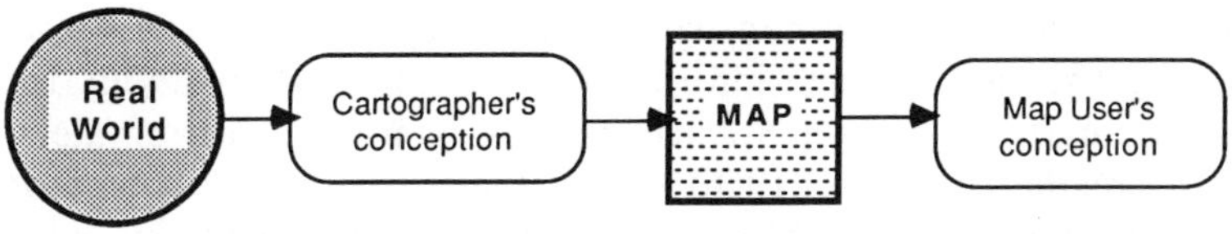

FIGURE 1 Simplified cartographic communication model (Source: Chrisman 1987)

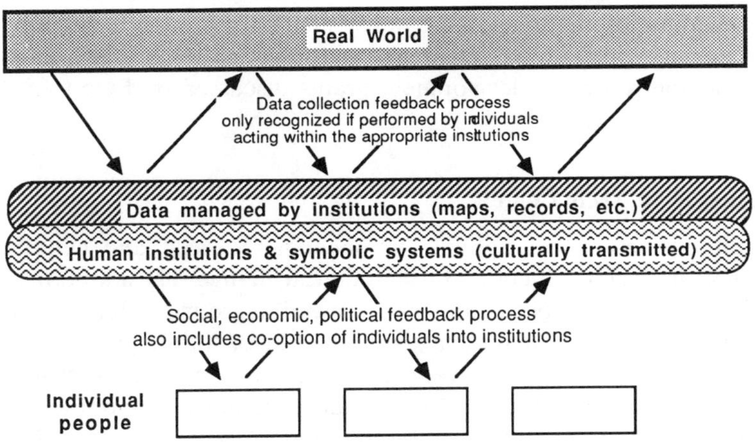

FIGURE 2 Cartographic communication carried out over time with cultural and institutional components (Source: Chrisman 1987)

Understanding Map Use and Benefits

A major contribution of the 'communication paradigm' in cartographic research was to move the research beyond the technology of making maps to consider the complete communication process. Research in the communication paradigm tended to focus on the individual map reader (Robinson and Petchenik 1976). Figure 1 presents the simplest form of the communication model.

This particular linkage between map maker and map reader is but one link in the full scope of the context of cartography. Studies in this field should expand to consider the use of maps and cartographic databases, not as simple messages transmitted by a map maker, but as a part of a larger social process. Chrisman (1987) portrayed the classical diagram of the communication model with an expansion to show the continuity of the organization over time (see Figure 2).

There have been some beginnings of such research. Muehrcke's (1986) '*Map Use*', though a text for introductory students, sets a broad stage for considering map use. Monmonier (1985) placed emphasis on the 'management' issues in cartography in the section quoted above. Recently, Nyerges (1991) and Chrisman (1990) have taken different, but complementary approaches on the issue. Taken together, these research efforts can be seen as an attempt to construct a 'meta-cartography'. This term derives from the use of meta-information as 'information about the information' in the computer science literature, not the use of meta-cartography made by Bunge (1966) where it meant a baldly geometric view of spatial relationships. Meta-cartography might include mapping the distribution of mapping (Szegö 1987), and understanding the social and cultural determinants of the differences so discovered. Research into the organization of mapping must study the diffusion of technology through normal professional channels and through such unexpected pathways as the source of development grants discussed by Eyre (1989) above.

Tools Engaged In Their Context

One result of meta-cartography is to understand social and cultural impediments in the introduction of technological change. The adoption of new technology will not occur on strictly scientific grounds. Real people in real institutions have to make changes in their lives. Although some new technology might offer overall advantages on technical criteria, these may not be the criteria actually used. Bureaucratic decisions to adopt a new technology or to alter the products of long standing may be influenced by many factors from the personal advancement of careers to overall cultural views. For example, just as technologies have life-cycles, the personnel in an agency have life-cycles as well. Most of the topographic mapping organizations required many newly trained photogrammetrists during the period 1935–55. A large number of these entered this discipline due to wartime emphasis on airphoto interpretation. These professionals rose in the ranks, according to normal bureaucratic procedures, so that at the time that the digital mapping technology emerged, the photogrammetrists tended to be in charge. Such events are often accepted as a fact of life, but it may have an influence. As new technology is developed, it should be engaged in the context it must enter.

There are additional limitations which inhibit the research process and the connection between research and software development. The academic base of early research in automated cartography had an effect which still continues (Chrisman 1988). Academic research may not reward the final integration of software to the point where it finally becomes usable by the typical end-user. For the future of the university this might be good or bad, depending on your perspective. The overall societal perspective might lead to different conclusions. A similar division occurs in the public sector, and even in the private sector. The rigours of the competitive market do not always reward strictly on the basis of technical merit.

Building Tools That Serve Goals

The next stage in meta-cartography involves connection from the spatial distributions to the reasons behind them. If Europe is mapped at an order of magnitude greater cost than any other region of the globe, do the citizens actually get something to show for it? Cartographic technologists must be prepared to defend their technology in the arena where resources are allocated. This will require a knowledge of the value of the information produced by our systems. A first step is the directly measurable 'benefits' which have been the subject of a few studies (Bernhardsen and Tveitval 1986, Epstein and Duchesneau 1984). Economic efficiency is certainly important to the future of the technology, but there are other measures of value which may override a pure cost–benefit calculation. Many of the spatial data collection activities of public organizations are not guided by a purely economic value. For example, the recording of property ownership and property valuation is justified in the USA as a measure to ensure equitable taxation. When a country moves away from a tax with some connection to land value (as the UK has recently in adopting a flat rate poll tax), the public interest in mapping will be reduced. Of course, due to bureaucratic inertia, the effects of taxation change may not become apparent immediately. The point is that spatial information may not be collected and processed for its inherent value (as either art or science). These activities which add up to form cartography fit into broader social purposes. Harley (1989) calls for a re-examination of maps in terms of the power relationships in society that the map reflects. Power is one part of the whole equation, but not the only one.

As a result, cartographic technology should advance with an eye on the actual problems posed by practice and by the social purposes for those practices. A new technology must be justified not for its scientific purity or even on its direct economic efficiency as a technology. The ultimate test is satisfying the purpose of having the information.

PRACTITIONER'S SUMMARY

While this paper has concentrated on the generalities, a practising cartographer has always been confronted with the practical politics of working inside some organization. Recent research shows that the research community is increasingly aware of the social and institutional context. It should be clear that technology does not succeed on its own merits. The context in which it is used is equally important. Thus, there is no single, universal formula for a successful implementation of new technology.

Over time, there should be less separation between academic conceptualization of cartography and the cartography actually practised. This development requires flow in both directions. Practitioners should share their findings with the research community through journals and proceedings.

PROJECTIVE SUMMARY

Cartographers, in developing new technology, will make choices with institutional and societal implications. Some of these choices may be made for seemingly technical reasons without considering the full range of broader implications. The traditional emphasis on the quality of the product and the economic efficiency of production will retain a place, but it should no longer remain the only criterion. Modern society cannot select projects simply on a simple measure such as benefit/cost ratios. There are criteria of balance between sectors of the public interest which are not measured on strictly numerical terms. For instance, many uses of modern spatial information systems involve complex decision-making circumstances where competing interests do not share a common set of values. Cartographers must learn to develop tools to bridge the gap and serve overall societal goals.

REFERENCES AND BIBLIOGRAPHY

Ayers E H (1984) Implementation of modern land records systems: Politics and institutional reform. *Proceedings Seminar on the MultiPurpose Cadastre*, IES Report 123. Madison, WI, University of Wisconsin: 289–99

Bernhardsen T, Tveitdal S (1986) Community benefit of digital spatial information. *Proceedings AUTO-CARTO London* **2**: 1–4.

Bie S (1984) Organizational needs for technological advancement. *Cartographica* **21** (2&3): 44–50

Blakemore M J, Harley J B (1980) Concepts in the history of cartography. *Cartographica* **17** (4): 1–120

Bunge W (1966) *Theoretical Geography*. Lund, Sweden, Gleerup

Chrisman N R (1987) Design of geographic information systems based on social and cultural goals. *Photogrammetric Engineering and Remote Sensing* **53** (10): 1367–70

Chrisman N R (1988) The risks of software innovation, a case study of the Harvard Lab. *The American Cartographer* **15**: 291–300

Chrisman N R (1990) Building a geography of geographic information: Placing GIS in cultural and historical context. (draft manuscript).

Chrisman N R, Niemann B J, Jr (1985) Alternative routes to a multipurpose cadastre: merging institutional and technical reasoning. *Proceedings AUTO-CARTO* 7: 84–94

Dangermond J, Freedman C (1984) Findings regarding a conceptual model of a municipal data base and implications for software design. *Proceedings Seminar on the MultiPurpose Cadastre*, IES Report 123. Madison, WI, University of Wisconsin: 12–49

Dickinson H J, Calkins H W (1988) The economic evaluation of implementing GIS. *International Journal of GIS* **2**: 307–27

Dickinson H J, Calkins H W (1990) Commenting on 'Concerning 'The economic evaluation of implementing GIS''. *International Journal of GIS* **4**: 211–12

Epstein E F, Duchesneau T D (1984) *The Use and Value of Geodetic Information.* Rockville, MD, Federal Geodetic Control Committee

Eyre L A (1989) JAMGIS, the first Jamaican Government comprehensive multi-data geographical information system. Achievements and problems. *International Journal of GIS* **3**: 363–72

Harley J B (1989) Deconstructing the map. *Cartographica* **26** (2): 1–20

Harley J B, Woodward D (ed.) (1987) *The History of Cartography*, Vol. 1. Chicago, University of Chicago Press

International Cartographic Association (1973) *Multilingual Dictionary of Technical Terms in Cartography.* Weisbaden, Franz Steiner Verlag

Laboratory for Computer Graphics (1973) *POLYVRT User's Manual.* Cambridge, MA, Harvard University

Mead D (1981) Statewide natural resource information systems: A status report. *Journal of Forestry* **79**: 369–90

Monmonier M S (1985) *Technological Transition in Cartography.* Madison, WI, University of Wisconsin Press

Muehrcke P C (1986) *Map Use* (2nd edition). Madison, WI, JP Publications

Nyerges T L (1991) Analytical map use. *Cartography and Geographic Information Systems* **18** (1): 11–22

Portner J, Niemann B J (1984) An overview of social aspects of land records modernization. *Proceedings Seminar on the MultiPurpose Cadastre*, IES Report 123. Madison, WI, University of Wisconsin: 94–108

Raisz E (1937) Charts of historical cartography. *Imago Mundi* **2**: 9–16

Robinson A H, Petchenik B B (1976) *The Nature of Maps: Essays towards Understanding Maps and Mapping.* Chicago, IL, University of Chicago Press

Robinette A (1984) Institutional innovations: Moving from academia to serving government needs. *Proceedings Seminar on the MultiPurpose Cadastre*, IES Report 123. Madison, WI, University of Wisconsin: 155–66

Scott A J, Storper M (ed.) (1986) *Production, Work and Territory: The Geographical Anatomy of Industrial Capitalism.* London, Allen & Unwin

Szegö J (1987) *Human Cartography.* Stockholm, Council for Building Research

Tomlinson R F (1974) The application of electronic computing methods and techniques to the storage, compilation and assessment of mapped data. Unpublished PhD thesis, University of London

Tomlinson R F, Petchenik B B (ed.) (1988) Reflections on the revolution: The transition from analogue to digital representations of space, 1958–1988. *The American Cartographer* **15** (3) special issue

Wilcox D L (1990) Concerning 'The economic evaluation of implementing GIS'. *International Journal of GIS* **4**: 203–10

Wunderlich G, Moyer D D (1984) Economic features of land information systems. *Proceedings Seminar on the MultiPurpose Cadastre*, IES Report 123. Madison, WI, University of Wisconsin: 183–202

LIST OF ABBREVIATIONS AND ACRONYMS

AA	Automobile Association
AI	Artificial Intelligence
ALI	Autofahrer Leit- und Informationssystem
AMS	Automated Mapping Systems
CAD/CAM	Computer-Aided Design/Computer-Aided Manufacturing
CARIN	Car Information & Navigation System
CDM	Conceptual Data Model
CD-ROM	Compact Disc/Read-Only Memory
CDS	Conceptual Data Scheme
CERCO	Comité Européen des Responsable del al Cartographie Officielle
CES	Cartographic Expert Systems
CMAS	Circular Map Accuracy Standard
CPU	Central Processing Unit
CRT	Cathode Ray Tube
DBMS	Database Management Systems
DCDSTF	Digital Cartographic Data Standards Task Force
DCW	Digital Chart of the World
DEM	Digital Elevation Model
DEMETER	Digital Electronic Mapping of European TERritories
DES	Department of Education & Science
DGI	Digital Geographic Information
DPS	Desktop Publishing Systems
EDI	Electronic Data Interchange
EF	Exchange Format
EVA	Elektronischer Verkehrlotse für Autofahrer

GDB	Geographic Database
GDBS	Geographic Database System
GDFR	Geographic Data File Release
GIS	Geographic Information System
HBDS	Hypergraph Based Data Structures
HIT	Human Interface Technologies
ICA	International Cartographic Association
IGN	Institut Géographic National
ISO	International Organization for Standardization
IT	Information Technology
ITASES	International Institute for Aerospace Survey & Earth Sciences
LISB	Leit- und Informationssystem Berlin
LTM	Long-term memory
MS-DOS	Microsoft Disk Operating System
NATO	North Atlantic Treaty Organization
NCGIA	National Center for Geographic Information and Analysis
NL	Natural language
NLF	Natural Language Front
NSF	National Science Foundation
PC	Personal Computer
PIES	Portable Interactive Edit Station
PQL	Pictorial Query Languages
QC	Quality Control
QL	Query Languages
QPE	Query-by-Pictorial Example
R&D	Research & Development
RDS	Radio Data Signal
SDA	Specification for Data Acquisition
SDC	Specification of Data Content
SDTS	Spatial Database Transfer Standard
SpQL	Spatial Query Languages
SQL	Structured Query Language
SRAS	Spatial Relations Acquisition Station
STM	Short-Term Memory
UDMS	Urban Datax Management System
USGS	United States Geological Survey
VDU	Visual Display Unit
VNS	Vehicle Navigation System

INDEX